International Mathematical Olympiad Volume I

T0178798

Anthem Science, Technology and Medicine

International Mathematical Olympiad Volume I

1959–1975

ISTVAN REIMAN

Anthem Press

Anthem Press
An imprint of Wimbledon Publishing Company
75-76 Blackfriars Road, London SE1 8HA
or
PO Box 9779, London SW19 7ZG
www.anthempress.com

This edition published by Anthem Press 2005.
First published by Typotex Ltd in Hungarian as
Nemzetközi Matematikal Diákolimpiák by I Reiman.

Translated by János Pataki, András Stipsitz & Csaba Szabó

British Library Cataloguing in Publication Data
A catalogue record for this book is available from the British Library.

Library of Congress Cataloguing in Publication Data
A catalogue record for this book has been requested.

1 3 5 7 9 10 8 6 4 2

ISBN 1 84331 197 6 (Hbk)
ISBN 1 84331 198 4 (Pbk)

Cover Illustration: Footprint Labs

Prelims typeset by Footprint Labs Ltd, London
www.footprintlabs.com

Printed in India

Preface

This three volume set contains the problems from the first forty-five IMO-s, from 1959 to 2004.

The chronicle of the IMO (International Mathematics Olympiad) starts with the initiative of the Romanian Mathematics and Physics Society: in July 1959 on the occasion of a celebration the Society invited high school students from the neighbouring countries to an international mathematical competition. The event proved to be such a success that the participants all agreed to go on with the enterprise. Ever since, this competition has taken place annually (except for 1980) and it has gradually transformed from the local contest of but a few countries into the most important and comprehensive international mathematical event for the young. Only seven nations were invited for the first IMO, while the number of participating countries was well beyond eighty for the last event; wherever mathematical education has reached a moderate level, sooner or later the country has turned up at the IMO.

The movement has had a significant impact on the mathematical education of several participating countries and also on the development of the gifted. The aim of a more proficient preparation for the IMO itself has launched the organization of national mathematical competitions in many countries involved. As the crucial component of successful participation, the preparation for the contest has enriched the publishing activity in several countries. Math-clubs have been formed on a large scale and periodicals have started. Even though the competition certainly brings up some pedagogical problems, if the educators regard the competitions not as ultimate aims, but as ways to introduce and endear pupils to mathematics, then their pedagogical benefit is undeniable.

The administration of the competitions has not changed that much; the larger scale has obviously necessitated certain modifications but the actual contest is more or less as it used to be. The participating countries are invited to delegate a group of up to six students who are attending high school at the year of the contest or had just finished their secondary school studies. Three problems are posed each day over two consecutive days and the students have to produce written solutions in their native tongue. There are two delegation leaders accompanying each team; one of their tasks is to provide an oral translation of their students' work into one of the official languages—by now this has been almost exclusively English—for a committee of mathematicians from the host country. Together with this group of coordinators they eventually settle the score the solutions are worth; the highest mark is seven points for each problem. The contestants are then ranked according to their total scores; the awarding of the prizes has been administered according to the following principle: half of the participants are given a prize: namely the proportion of the gold, silver and bronze medals is $1:2:3$ respectively.

The occasional professional problems are handled by the international jury formed by the leaders of the participating delegations; their most important and difficult task is to select the six problems for the actual contest, to formulate their official text and to prepare rough marking schemes for each of them. The organizers ask for proposals from the participating countries well in advance; in due course they produce a list of approximately twenty to twenty five problems from those suggested and the jury selects the final six from this supply.

There are almost two hundred problems in this three volume set and they provide a full image of the challenge the students had to cope with during these forty years. One cannot claim that every single one of them is a pearl of mathematics but their overwhelming majority is interesting and rewarding; together they more or less cover the usual syllabus-chapters of elementary mathematics. When selecting, the jury usually tries to choose from the intersection of the respective curricula of the participating countries; considering that there are more than eighty of them this is not an easy job, if not impossible. The reader might notice that there are no problems at all from the theory of probability, for example, and complex numbers hardly show up.

From the retrospect of more than forty years one can certainly conclude that the IMO movement has had a significant role in the history of the second half of twentieth century mathematics. There are quite a few highly ranked mathematicians who started their career at an IMO; even at this point, however, we have to emphasize, that an eventual fiasco at the IMO or any other mathematical contest whatsoever usually has no implications at all about the mathematical potential of a well prepared student.

A careful reader will certainly realize that quite a few problems in this book are in fact simplifications or particular cases of more profound mathematical results; apart from the intellectual satisfaction of actually solving these problems, the discovery of this mathematical background and the knowledge gained from it can be the ultimate benefits of a high level study of this book.

At the end of the book we included a Glossary of Theorems (and their proofs) we used in the book and we refer to them by their numbers enclosed in brackets, e. g. [6].

István Reiman

International Mathematical Olympiad

Problems

1959.

1959/1. Prove that the expression $\dfrac{21n+4}{14n+3}$ is irreducible for every positive integer n.

1959/2. Determine the real solutions of the following equations.

a) $\sqrt{x+\sqrt{2x-1}}+\sqrt{x-\sqrt{2x-1}}=\sqrt{2}$,

b) $\sqrt{x+\sqrt{2x-1}}+\sqrt{x-\sqrt{2x-1}}=1$,

c) $\sqrt{x+\sqrt{2x-1}}+\sqrt{x-\sqrt{2x-1}}=2$.

1959/3. Suppose that x satisfies

$$(1) \qquad a\cos^2 x+b\cos x+c=0.$$

Form a quadratic equation in $\cos 2x$ whose roots are the same values of x. Apply your result for the special case $a=4$, $b=2$, $c=-1$.

1959/4. Construct a right triangle with a given hypotenuse, if we know that the median corresponding to the hypotenuse is equal to the geometric mean of the two adjacent sides.

1959/5. A point M is moving on the interval AB. Squares $AMCD$ and $BMEF$ over the subintervals AM and MB (on the same side of the line AB) are constructed The circumcircles of these squares intersect each other in the points M and N. Verify that the lines AE and BC pass through N. Show that for any choice of M the line MN passes through a fixed point. Find the locus of the midpoints of the intervals joining the centres of the two squares.

1959/6. The planes P and Q intersect in the line p. A and C are points of the planes P and Q respectively, neither of them is on p. Construct that symmetric trapezium $ABCD$ (with $AB \parallel CD$) for which its vertices B and D are on the planes P and Q and $ABCD$ admits an incircle.

1960

1960/1. Determine all three digit numbers which are equal to 11 times the sum of the squares of their digits.

1960/2. Determine the real solutions of the inequality

$$\frac{4x^2}{\left(1 \perp \sqrt{1+2x}\right)^2} < 2x+9.$$

1960/3. The hypotenuse $BC = a$ of the right triangle ABC has been divided into n equal intervals with n an odd integer. Let h denote the altitude corresponding to the hypotenuse; and the central interval subtends an angle α at A. Show that

$$\tan \alpha = \frac{4nh}{(n^2 \perp 1)a}.$$

1960/4. Construct the triangle ABC if its two altitudes m_a and m_b (corresponding to the vertices A and B) and the median corresponding to A is given.

1960/5. For a given cube $ABCDA'B'C'D'$ let X be a point of the face diagonal AC, and Y a point of $B'D'$.

a) Find the locus of the midpoints of the intervals XY for all possible choices of X and Y.

b) Consider the point $Z \in XY$ satisfying the equality $ZY = 2XZ$ and determine the locus of these points for all choices of X and Y.

1960/6. A cone of revolution has an inscribed sphere tangent to the base of the cone (and to the sloping surface of the cone). A cylinder is circumscribed about the sphere so that its base lies in the base of the cone. Let V_1 and V_2 denote the volume of the cone and the resulting cylinder, respectively.

a) Show that V_1 is not equal to V_2.

b) Determine the smallest possible value of $k = \dfrac{V_1}{V_2}$, and for this minimal k construct the half angle of the symmetric cone.

1960/7. The parallel sides of a symmetric trapezium are of length a and b, while its altitude is m.

a) Construct the point P on the symmetry axis of the trapezium which is on the Thales circles over the legs of the trapezium.

b) Determine the distance of P from one of the parallel sides.

c) Under what assumption does such P exist?

1961

1961/1. Solve the following system of equations for x, y and z:

(1)
$$x + y + z = a,$$
(2)
$$x^2 + y^2 + z^2 = b^2,$$
(3)
$$xy = z^2,$$

where a and b are given real numbers. What conditions must a and b satisfy for x, y and z to be all positive and distinct?

1961/2. Let a, b and c be the sides of a given triangle while t is its area. Show that

(1)
$$a^2 + b^2 + c^2 \geq 4t\sqrt{3}$$

When does equality hold?

1961/3. Solve the equation

$$\cos^n x \perp \sin^n x = 1$$

where n is a positive integer.

1961/4. P is a point inside the triangle $P_1 P_2 P_3$. The intersections of the lines $P_1 P$, $P_2 P$ and $P_3 P$ with the opposite sides are denoted by Q_1, Q_2 and Q_3, respectively. Show that there is one among the ratios

$$\frac{P_1 P}{PQ_1}, \quad \frac{P_2 P}{PQ_2}, \quad \frac{P_3 P}{PQ_3}$$

which is not less, and one which is not more than 2.

1961/5. Construct a triangle ABC if the length of the two sides $AC = b$ and $AB = c$ and the acute angle $AMB = \omega$ is given — here M is the midpoint of BC. Show also that the problem admits a solution if and only if

$$b \tan \frac{\omega}{2} \leq c < b.$$

1961/6. Let ε be a given plane and A, B, C three non-collinear points on one side of ε. Suppose furthermore that the plane determined by these points is parallel to ε. Let A', B' and C' be three arbitrary points on ε. The midpoints of AA', BB' and CC' are denoted by L, M and N respectively. The centre of gravity of the triangle LMN is denoted by G. (Those triples A', B', C' for which L, M, N do not form a triangle, are disregarded.) Determine the locus of G for any possible choice of the triple A', B', C' on the plane ε.

1962

1962/1. Determine the smallest possible positive integer x whose last decimal digit is 6, and if we erase this last 6 and put it in front of the remaining digits, we get four times x.

1962/2. Determine all real x satisfying

$$\sqrt{3-x}-\sqrt{x+1}>\frac{1}{2}.$$

1962/3. The cube $ABCDA'B'C'D'$ with upper face $ABCD$ and lower face $A'B'C'D'$ ($AA' \parallel BB' \parallel CC' \parallel DD'$) is given. A point X runs along the perimeter of $ABCD$ (in the direction given by the above order) with constant speed, while a point Y does the same (with equal speed) along the perimeter of the square $B'C'CB$. X and Y start in the same instant from A and B', respectively. Determine the locus of the midpoint Z of the interval XY.

1962/4. Solve the following equation:
$$\cos^2 x + \cos^2 2x + \cos^2 3x = 1.$$

1962/5. Three distinct points A, B and C on a circle k are given. Construct the point D on the circle for which the quadrilateral $ABCD$ admits an incircle.

1962/6. Let R and r denote the radii of the circumcircle and the incircle of an isosceles triangle. Show that the distance d between the centres of the two circles is

(1) $$d = \sqrt{R(R-2r)}.$$

1962/7. There are five spheres which are tangent to all extended edges of a tetrahedron $SABC$. Show that

a) $SABC$ is a regular tetrahedron;

b) conversely: a regular tetrahedron admits five spheres with the properties described above.

1963

1963/1. Determine the real solutions of the following equality (p denotes a real parameter)

$$\sqrt{x^2-p}+2\sqrt{x^2-1}=x.$$

1963/2. Given a point A and a segment BC in the 3-dimensional space, determine the locus of those points, P, for which the angle $\angle APX$ is a right angle for some X on the segment BC.

1963/3. Consider a convex n-gon with equal angles and with consecutive sides $a_1, a_2, \ldots, a_n$ satisfying

(1)
$$a_1 \geq a_2 \geq \ldots \geq a_n.$$

Show that under the above conditions we have

(2)
$$a_1 = a_2 = \ldots = a_n.$$

1963/4. Determine the values x_1, x_2, x_3, x_4, x_5 satisfying

(1)
$$x_5 + x_2 = yx_1,$$

(2)
$$x_1 + x_3 = yx_2,$$

(3)
$$x_2 + x_4 = yx_3,$$

(4)
$$x_3 + x_5 = yx_4,$$

(5)
$$x_4 + x_1 = yx_5$$

where y is a given parameter.

1963/5. Show that

(1)
$$\cos \frac{\pi}{7} \perp \cos \frac{2\pi}{7} + \cos \frac{3\pi}{7} = \frac{1}{2}.$$

1963/6. Five students, A, B, C, D and E were placed 1 to 5 in a contest. Someone made the initial guess that the final result would be the order $ABCDE$, but — as it turned out — this person was wrong on the final position of all the contestants; moreover no two students predicted to finish consecutively did so. A second person guessed $DAECB$, which was much better, since exactly two contestants finished in the place predicted, and two disjoint pairs predicted to finish consecutively did so. Determine the outcome of the contest.

1964

1964/1. a) Find all positive integers n for which 7 divides $2^n \perp 1$.

b) Show that there is no positive integer n for which 7 divides $2^n + 1$.

1964/2. Let a, b and c denote the lengths of the sides of a triangle. Show that

(1)
$$a^2(\perp a + b + c) + b^2(a \perp b + c) + c^2(a + b \perp c) \leq 3abc.$$

1964/3. Let a, b, c denote the lengths of the sides of the triangle ABC. Tangents to the inscribed circle are constructed parallel to the sides. Each tangent forms a triangle with the other two sides of the triangle, and a circle is inscribed in each of these three triangles. Find the total area of all four inscribed circles.

Consider the tangents of the incircle which are parallel to the sides. These tangents give rise to three subtriangles of ABC, consider the incircles of these subtriangles. Determine the sum of the areas of the four incircles.

1964/4. Each pair from 17 scientists exchange letters on one of three topics. Prove that there are at least three scientists who write to each other on the same topic.

1964/5. Five points on the plane are situated so that no two of the lines joining a pair of points are coincident, parallel or perpendicular. Through each point lines are drawn perpendicular to each of the lines through two of the other four points. Give the best possible upper bound for the number of intersection points of these orthogonals, disregarding the given 5 points.

1964/6. Let $ABCD$ be a given tetrahedron and D_1 the centroid of the face ABC. The parallels to DD_1 passing through the vertices A, B and C intersect the opposite faces in A_1, B_1 and C_1, respectively.

a) Show that the volume of $ABCD$ is one-third the volume of $A_1B_1C_1D_1$.

b) Is the result valid for any choice of D_1 in the interior of ABC?

1965

1965/1. Find all x in the interval $[0, 2\pi]$ which satisfy

(1) $$2\cos x \le |\sqrt{1+\sin 2x} \perp \sqrt{1 \perp \sin 2x}| \le \sqrt{2}.$$

1965/2. The coefficients of the system of equations

$$a_{11}x_1 + a_{12}x_2 + a_{13}x_3 = 0,$$
$$a_{21}x_1 + a_{22}x_2 + a_{23}x_3 = 0,$$
$$a_{31}x_1 + a_{32}x_2 + a_{33}x_3 = 0$$

are subject to the following constraints:

a) a_{11}, a_{22} and a_{33} are all positive,

b) all other coefficients are negative,

c) the sum of coefficients in each equation is positive.

Verify that the only solution of the system is

$$x_1 = x_2 = x_3 = 0.$$

1965/3. The length of the edge AB in the tetrahedron $ABCD$ is a, while the length of CD is b. The distance between the skew lines AB and CD is d, the angle determined by them is ω. The tetrahedron is divided into two parts by a plane ε parallel to AB and CD. We also know that k times the distance between AB and ε equals the distance between CD and ε. Determine the ratio of the volumes of the parts of the tetrahedron.

1965/4. Find all sets of four real numbers x_1, x_2, x_3, x_4 such that the sum of any one and the product of the other three is 2.

1965/5. The triangle OAB has angle $\angle AOB$ acute. M is an arbitrary point in OAB different from O. The points P and Q are the feet of the perpendiculars from M to OA and OB, respectively. Determine the locus of the orthocentre H of the triangle OPQ if M is

a) on AB;

b) in the interior of OAB.

1965/6. For $n \geq 3$ points in the plane denote the maximal distance of pairs of points by d. Prove that at most n pairs of points are of distance d apart.

1966

1966/1. Problems A, B and C have been posed in a mathematical contest. 25 competitors solved at least one of the three. Amongst those who did not solve A, twice as many solved B as C. The number of competitors solving only A was one more than the number of competitors solving A and at least one other problem. The number of competitors solving A equalled the number solving just B plus the number of competitors solving just C. How many competitors solved just B?

1966/2. Let a, b, c denote the sides of a triangle, while the opposite angles are denoted by α, β, γ. Prove that if

(1) $$a + b = \tan \frac{\gamma}{2}(a \tan \alpha + b \tan \beta)$$

then the triangle is isosceles.

1966/3. Prove that a point in the space has the smallest sum of distances to vertices of a regular tetrahedron if and only if it is the centre of the tetrahedron.

1966/4. Prove that

(1) $$\frac{1}{\sin 2x} + \frac{1}{\sin 4x} + \ldots + \frac{1}{\sin 2^n x} = \cot x \perp \cot 2^n x.$$

for any positive integer n and any real x (with $x \neq \frac{\lambda \pi}{2^k}$, $k = 0, 1, 2, \ldots, n$ and λ an arbitrary integer).

1966/5. Solve the system of equations
$$|a_1 \perp a_2|x_2 + |a_1 \perp a_3|x_3 + |a_1 \perp a_4|x_4 = 1,$$
$$|a_2 \perp a_1|x_1 \qquad + |a_2 \perp a_3|x_3 + |a_2 \perp a_4|x_4 = 1,$$
$$|a_3 \perp a_1|x_1 + |a_3 \perp a_2|x_2 \qquad + |a_3 \perp a_4|x_4 = 1,$$
$$|a_4 \perp a_1|x_1 + |a_4 \perp a_2|x_2 + |a_4 \perp a_3|x_3 \qquad = 1,$$
where a_1, a_2, a_3, a_4 denote four distinct reals.

1966/6. Take any points K, L, M on the sides AB, BC, CA of the triangle ABC. Prove that at least one of the triangles MAL, KBM and LCK has area at most fourth the area of ABC.

1967

1967/1. The parallelogram $ABCD$ has $AB = a$, $AD = 1$, angle $\angle DAB = \alpha$ and the triangle ABD is acute. Prove that the circles K_A, K_B, K_C and K_D of radius 1 centered at A B, C and D cover the parallelogram if and only if

(1) $$a \leq \cos\alpha + \sqrt{3}\sin\alpha.$$

1967/2. Prove that a tetrahedron with just one edge of length greater than 1 has volume at most $\dfrac{1}{8}$.

1967/3. Let k, m, n be positive integers such that $m + k + 1$ is a prime greater than $(n + 1)$ and let $c_s = s(s + 1)$ ($s = 1, 2, \ldots$). Prove that the product

(1) $$(c_{m+1} \perp c_k)(c_{m+2} \perp c_k) \cdot \ldots \cdot (c_{m+n} \perp c_k)$$

is divisible by

(2) $$c_1 c_2 \ldots c_n.$$

1967/4. $A_0 B_0 C_0$ and $A'B'C'$ are given acute triangles. Construct the triangle ABC with the largest possible area which is circumscribed around $A_0 B_0 C_0$ (i.e., AB contains C_0, BC contains A_0 and CA contains B_0) and is similar to $A'B'C'$ (A, B, C correspond to A', B' and C').

1967/5. Consider the sequence $\{c_n\}$ given as

$$c_1 = a_1 + a_2 + \ldots + a_8$$
$$c_2 = a_1^2 + a_2^2 + \ldots + a_8^2$$
$$\vdots$$
$$c_n = a_1^n + a_2^n + \ldots + a_8^n,$$
$$\vdots$$

for $a_1, a_2, \ldots, a_8$ reals, not all zero.

Given that an infinite number of $\{c_n\}$ is zero, find all n for which $c_n = 0$.

1967/6. In a sports contest a total of m medals were awarded over $n > 1$ days. On the first day one medal and $\dfrac{1}{7}$ of the remaining medals were awarded. On the second day two medal and $\dfrac{1}{7}$ of the remaining medals were awarded, and so on. On the last day the remaining n medals were awarded. How many medals and over how many days were awarded?

1968

1968/1. Show that there is a unique triangle whose side lengths are consecutive integers and one of whose angles is twice another.

1968/2. Find all positive integers x for which

(1) $$p(x) = x^2 \perp 10x \perp 22,$$

where $p(x)$ is the product of the decimal digits of x.

1968/3. Prove that the system

$$ax_1^2 + bx_1 + c = x_2,$$
$$ax_2^2 + bx_2 + c = x_3,$$
(1)
$$\dots\dots\dots\dots\dots$$
$$ax_n^2 + bx_n + c = x_1$$

(a, b, c are real with $a \neq 0$)
  I. has no real solution if $(b \perp 1)^2 \perp 4ac < 0$;
 II. has one real solution if $(b \perp 1)^2 \perp 4ac = 0$;
III. and has more than one real solution once $(b \perp 1)^2 \perp 4ac > 0$.

1968/4. Prove that every tetrahedron has a vertex whose three edges have the right lengths to form a triangle.

1968/5. Let f be a real-valued function defined for all real numbers, such that for some $a > 0$ it satisfies

(1) $$f(x + a) = \frac{1}{2} + \sqrt{f(x) \perp (f(x))^2}.$$

  I. Prove that f is periodic, i.e., there exists a positive real b such that

$$f(x + b) = f(x)$$

holds for every x.
  II. Give an example of such a non-constant f for $a = 1$.

1968/6. Let $[x]$ denote the greatest integer not larger than x (the "integer part" of x). For every positive integer n evaluate the sum

(1) $$\left[\frac{n+1}{2}\right] + \left[\frac{n+2}{2^2}\right] + \dots + \left[\frac{n+2^k}{2^{k+1}}\right] + \dots$$

1969

1969/1. Prove that there are infinitely many positive integers a such that

$$z = n^4 + a$$

is not a prime for any positive integer n.

1969/2. Let

(1) $\qquad f(x) = \cos(a_1 + x) + \dfrac{\cos(a_2 + x)}{2} + \dfrac{\cos(a_3 + x)}{2^2} + \ldots + \dfrac{\cos(a_n + x)}{2^{n-1}},$

where $a_1, a_2, \ldots, a_n$ are real constants and x is a real variable. Prove that if $f(x_1) = f(x_2) = 0$ then $x_2 - x_1 = m\pi$ for some integer m.

1969/3. For each $k = 1, 2, 3, 4, 5$ find necessary and sufficient conditions on $a > 0$ such that there exists a tetrahedron with k edges of length a and $(6 - k)$ edges of length 1.

1969/4. Let C be an interior point of the semicircle k over AB and D is the foot of the perpendicular from C to AB. The circle k_1 is the incircle of ABC, the circle k_2 touches CD, DA and k while k_3 touches CD, DB and k. Show that k_1, k_2 and k_3 have another common tangent apart from AB.

1969/5. Given $n > 4$ points on the plane (no three collinear), prove that there are at least $\dbinom{n-3}{2}$ convex quadrilaterals with vertices amongst the given points.

1969/6. For given real numbers $x_1, x_2, y_1, y_2, z_1, z_2$ satisfying $x_1 > 0$, $x_2 > 0$, $x_1 y_1 - z_1^2 > 0$ and $x_2 y_2 - z_2^2 > 0$, prove that

(1) $\qquad \dfrac{8}{(x_1 + x_2)(y_1 + y_2) - (z_1 + z_2)^2} \leq \dfrac{1}{x_1 y_1 - z_1^2} + \dfrac{1}{x_2 y_2 - z_2^2}.$

Give necessary and sufficient conditions for equality.

1970

1970/1. M is a point on the side AB of the triangle ABC. Let r_1, r_2 and r denote the radii of the incircles of AMC, BMC and ABC, respectively. ϱ_1, ϱ_2 and ϱ stands for the radii of the excircles of the triangles AMC, BMC and ABC (corresponding to sides AM, BM and AB), respectively. Prove that $\dfrac{r_1}{\varrho_1} \cdot \dfrac{r_2}{\varrho_2} = \dfrac{r}{\varrho}.$

1970/2. Real numbers x_i $(i = 0, 1, \ldots, n)$ with $0 \leq x_i < b$ and $x_n > 0$, $x_{n-1} > 0$ are given. If $x_n x_{n-1} \ldots x_1 x_0$ represents the number A_n base a and

B_n base b whilst $x_{n-1} \ldots x_1 x_0$ represents A_{n-1} base a and B_{n-1} base b, then prove that $a > b$ holds if and only if

(1) $$\frac{A_{n-1}}{A_n} < \frac{B_{n-1}}{B_n}$$

1970/3. The real numbers a_0, a_1, a_2, ..., a_n, ... satisfy

(1) $$1 = a_0 \leq a_1 \leq a_2 \leq \ldots \leq a_n \leq \ldots$$

We define the sequence b_1, b_2, ..., b_n, ... as

(2) $$b_n = \sum_{k=1}^{n} \left(1 \perp \frac{a_{k-1}}{a_k}\right) \frac{1}{\sqrt{a_k}}.$$

I. Prove that $0 \leq b_n < 2$ holds for all n.

II. Given c satisfying $0 \leq c < 2$, prove that we can find a_0, a_1, ..., a_n, ... (satisfying (1)) so that infinitely many of the corresponding b_n are greater than c.

1970/4. Find all positive integers n such that the set $\{n, n+1, n+2, n+3, n+4, n+5\}$ can be partitioned into two subsets so that the product of the numbers in each subset is equal.

1970/5. In the tetrahedron $ABCD$ the angle $\angle BDC$ is a right angle and the foot of the perpendicular from D to ABC is the intersection of the altitudes of ABC. Prove that

(1) $$(AB + BC + CA)^2 \leq 6(AD^2 + BD^2 + CD^2).$$

When do we have equality?

1970/6. Given 100 coplanar points (no three collinear), consider all triangles with vertices among the given points and prove that at most 70 % of these triangles have all angles acute.

1971

1971/1. Show that the following statement is true for $n = 3$ and 5 and false for all other $n > 2$:

"For any real numbers a_1, a_2, ..., a_n the inequality

$$(a_1 \perp a_2)(a_1 \perp a_3) \ldots (a_1 \perp a_n) + (a_2 \perp a_1)(a_2 \perp a_3) \ldots (a_2 \perp a_n) + \ldots$$
$$\ldots + (a_n \perp a_1)(a_n \perp a_2) \ldots (a_n \perp a_{n-1}) \geq 0.$$

holds".

1971/2. Let P_1 be a convex polyhedron with vertices A_1, A_2, ..., A_9. Let P_i be the polyhedron obtained from P_1 by a translation that moves A_1 into A_i ($i = 2, 3, ..., 9$). Show that at least two of the polyhedra P_1, P_2, ..., P_9 have an interior common point.

1971/3. Prove that we can find an infinite set of positive integers of the form $\{2^n \perp 3\}$ (where n is a positive integer) every pair of which are relatively prime.

1971/4. All faces of the tetrahedron $ABCD$ are acute triangles. Let X, Y, Z and T be points in the interiors of the segments AB, BC, CD and DA, respectively; and consider the closed path $XYZTX$.

a) If

(1) $$\angle DAB + \angle BCD \neq \angle ABC + \angle CDA,$$

then prove that none of the closed paths $XYZTX$ has minimal length.

b) If

(2) $$\angle DAB + \angle BCD = \angle ABC + \angle CDA,$$

then prove that there are infinitely many shortest paths $XYZTX$, each with length

$$2AC \sin \frac{\alpha}{2},$$

where $\alpha = \angle BAC + \angle CAD + \angle DAB$.

1971/5. Prove that for every positive integer m we can find a finite set of points S in the plane such that for any point A of S, there are exactly m points in S at unit distance from A.

1971/6. Let $A =$

$$
\begin{matrix}
a_{11} & a_{12} & \cdots & a_{1n} \\
a_{21} & a_{22} & \cdots & a_{2n} \\
\vdots & & & \\
a_{n1} & a_{n2} & \cdots & a_{nn}
\end{matrix}
$$

be a square matrix with all a_{ij} nonnegative integers. For each i, j with $a_{ij} = 0$ we have

(1) $$a_{i1} + a_{i2} + \ldots + a_{in} + a_{1j} + a_{2j} + \ldots + a_{nj} \geq n.$$

Prove that the sum of all the elements in the matrix is at least

$$\frac{n^2}{2}.$$

1972

1972/1. Given any set of ten distinct numbers in the range 10, 11, ..., 99, prove that we can always find two disjoint subsets with the same sum.

1972/2. Given $n > 4$, prove that every cyclic quadrilateral can be dissected into n cyclic quadrilaterals.

1972/3. Let m and n be nonnegative integers. Prove that
$$\frac{(2m)!(2n)!}{m!n!(m+n)!}$$
is an integer. (According to our conventions $0! = 1$).

1972/4. Find all positive real solutions $(x_1, x_2, x_3, x_4, x_5)$ of the system

(1) $\qquad (x_1^2 \perp x_3 x_5)(x_2^2 \perp x_3 x_5) \leq 0,$

(2) $\qquad (x_2^2 \perp x_4 x_1)(x_3^2 \perp x_4 x_1) \leq 0,$

(3) $\qquad (x_3^2 \perp x_5 x_2)(x_4^2 \perp x_5 x_2) \leq 0,$

(4) $\qquad (x_4^2 \perp x_1 x_3)(x_5^2 \perp x_1 x_3) \leq 0,$

(5) $\qquad (x_5^2 \perp x_2 x_4)(x_1^2 \perp x_2 x_4) \leq 0.$

1972/5. Let f and g be two real valued functions defined on the real line satisfying

(1) $\qquad f(x+y) + f(x \perp y) = 2f(x)g(y).$

Suppose that $f(x)$ is not identically zero and $|f(x)| \leq 1$ for all x. Show that $|g(y)| \leq 1$ for all y.

1972/6. Given four parallel planes, prove that there exists a regular tetrahedron with a vertex on each plane.

1973

1973/1. Let $\overrightarrow{OP_1}, \overrightarrow{OP_2}, \ldots, \overrightarrow{OP_n}$ be unit vectors in a plane. $P_1, P_2, \ldots, P_n$ all lie on the same side of a line through O. Prove that if n is odd, then
$$|\overrightarrow{OP_1} + \overrightarrow{OP_2} + \ldots + \overrightarrow{OP_n}| \geq 1,$$
where $|\overrightarrow{OM}|$ denotes the length of the vector $\overrightarrow{OM}$.

1973/2. Can we find a finite set of non-coplanar points, such that given any two points, A and B, there are two others, C and D, with the lines AB and CD parallel and distinct?

1973/3. Let a and b be real numbers for which the equation $x^4 + ax^3 + bx^2 + ax + 1 = 0$ has at least one real solution. Find the least possible value of $a^2 + b^2$.

1973/4. A soldier needs to sweep a region of the shape of an equilateral triangle for mines. The detector has an effective radius equal to half the altitude of the triangle. He starts at a vertex of the triangle. What path should he follow in order to travel the least distance and still sweep the whole region?

1973/5. Let G be a set of non-constant functions f. Each f is defined on the real line and has the form $f(x) = ax + b$ for some real a, b.

a) If $f, g \in G$, then so is $g \circ f \in G$, where $g \circ f(x) = g(f(x))$.

b) If f is in G, then so is the inverse f^{-1}.

$$f^{-1}(x) = \frac{x}{a} \perp \frac{b}{a}.$$

c) Every f in G has a fixed point. In other words we can find x_f such that

$$f(x_f) = x_f.$$

Prove that all the functions in G have a common fixed point.

1973/6. Let $a_1, a_2, \ldots, a_n$ be positive reals, and q satisfies $0 < q < 1$. Find $b_1, b_2, \ldots, b_n$ such that:

a) $a_k < b_k$, $(k = 1, 2, \ldots, n)$;

b) $q < \dfrac{b_{k+1}}{b_k} < \dfrac{1}{q}$, $(k = 1, 2, \ldots, n \perp 1)$;

c) $b_1 + b_2 + \ldots + b_n < \dfrac{1+q}{1 \perp q} (a_1 + a_2 + \ldots + a_n)$.

1974

1974/1. Three players A, B, C play the following game: There are three cards each with a different positive integer p, q and r where $p < q < r$. In each round the cards are randomly dealt to the players and each receives the number of counters on his card.

After two or more rounds, A has received 20, B 10 and C 9 counters. In the last round B received the largest number of counters.

Who received q counters in the first round?

1974/2. Let A, B and C denote the vertices of a triangle. Prove that there is a point D on the side AB of the triangle ABC, such that CD is the geometric mean of AD and DB if and only if

(1)
$$\sin A \sin B \le \sin^2 \frac{C}{2}.$$

1974/3. Prove that

$$\sum_{k=0}^{n} \binom{2n+1}{2k+1} \cdot 2^{3k}$$

is not divisible by 5 for any nonnegative integer n.

1974/4. An 8×8 chessboard is divided into p disjoint rectangles (along the lines between squares), so that

a) each rectangle has the same number of white squares as black squares

b) If a_i denotes the number of white squares in the i-th rectangle, then $a_1 < < a_2 < \ldots < a_p$ holds.

Find the maximum possible value of p and all possible $a_1, a_2, \ldots, a_p$ sequences.

1974/5. Determine all possible values of

$$S = \frac{a}{d+a+b} + \frac{b}{a+b+c} + \frac{c}{b+c+d} + \frac{d}{c+d+a}$$

for arbitrary positive reals a, b, c, d.

1974/6. Let $P(x)$ be a non constant polynomial with integer coefficients. Let n be the number of distinct integers k, where $(P(k))^2 = 1$. Prove that

(1)
$$n(P) \perp \deg(P) \le 2,$$

where $\deg P$ denotes the degree of $P(x)$.

1975

1975/1. Let $x_1 \ge x_2 \ge \ldots \ge x_n$ and $y_1 \ge y_2 \ge \ldots \ge y_n$ be real numbers. Prove that if $\{z_i\}$ is any permutation of the $\{y_i\}$, then:

(1)
$$\sum_{i=1}^{n}(x_i \perp y_i)^2 \le \sum_{i=1}^{n}(x_i \perp z_i)^2.$$

1975/2. Let $a_1 < a_2 < a_3 < \ldots$ be an infinite sequence of positive integers.

Prove that there are infinitely many elements of this sequence that can be written in the form

$$a_m = x a_p + y a_q,$$

with x, y positive integers and $p \ne q$.

1975/3. Given any triangle ABC, construct external triangles ABR, BCP, CAQ on the sides, so that

$$\angle PBC = \angle CAQ = 45°$$
$$\angle BCP = \angle QCA = 30°$$
$$\angle ABR = \angle BAR = 15°.$$

Prove that $\angle QRP = 90°$ and $QR = RP$.

1975/4. Let A denote the sum of the decimal digits of 4444^{4444}, and B be the sum of the decimal digits of A. Find the sum of the decimal digits of B.

1975/5. Can you find 1975 points on the circumference of a unit circle such that the distance between each pair is rational?

1975/6. Find all polynomials $P(x, y)$ in two variables such that:

I. for every real numbers t, x, y, $P(tx, ty) = t^n P(x, y)$, where n is a positive integer, i.e. P is a homogenous polynomial of degree n.

II. For every real a, b, c, $P(a+b, c) + P(b+c, a) + P(c+a, b) = 0$.

III. $P(1, 0) = 1$.

Solutions

1959.

1959/1. *Prove that the expression* $\dfrac{21n+4}{14n+3}$ *is irreducible for every positive integer* n.

First solution. If d divides two integers, then it divides their multiples, their sum and difference. Since

$$3(14n+3) \perp 2(21n+4) = 1,$$

a common divisor of the numerator and the denominator also divides 1; for this reason the fraction cannot be simplified.

Second solution. Notice that if $\dfrac{a}{b} = e + \dfrac{c}{b}$ (a, b, c, e integers) then $c = a \perp \perp be$ and $a = c + be$, consequently the common divisors of a and b also divide c; and similarly the common divisors of c and b divide a. For this reason $\dfrac{a}{b}$ can be simplified if and only if $\dfrac{c}{b}$ can be simplified; and the same holds for $\dfrac{a}{b}$ and $\dfrac{b}{a}$. Since

$$\frac{21n+4}{14n+3} = 1 + \frac{7n+1}{14n+3}$$

and

$$\frac{14n+3}{7n+1} = 2 + \frac{1}{7n+1},$$

and the last term cannot be simplified, the same is true for our original expression.

1959/2. *Determine the real solutions of the following equations.*

a) $\sqrt{x + \sqrt{2x \perp 1}} + \sqrt{x \perp \sqrt{2x \perp 1}} = \sqrt{2},$

b) $\sqrt{x + \sqrt{2x \perp 1}} + \sqrt{x \perp \sqrt{2x \perp 1}} = 1,$

c) $\sqrt{x + \sqrt{2x \perp 1}} + \sqrt{x \perp \sqrt{2x \perp 1}} = 2.$

Solution. First of all notice that

$$x + \sqrt{2x \perp 1} = \frac{1}{2}\left(1 + \sqrt{2x \perp 1}\right)^2 \quad \text{and} \quad x \perp \sqrt{2x \perp 1} = \frac{1}{2}\left(1 \perp \sqrt{2x \perp 1}\right)^2.$$

(Here we have to assume $x \geq \frac{1}{2}$.) Let B denote the expression on the left hand side of the equation:

$$B = \frac{1}{\sqrt{2}}\left(\left|1+\sqrt{2x-1}\right|+\left|1-\sqrt{2x-1}\right|\right).$$

Since for $x \geq \frac{1}{2}$ we have $1+\sqrt{2x-1}>0$, the absolute value sign can be omitted. The second expression is nonnegative if

(1) $$1 \geq \sqrt{2x-1}.$$

Since both sides of (1) are nonnegative, this inequality is equivalent to

$$1 \geq 2x-1, \quad \text{i.e.} \quad x \leq 1.$$

According to this,

$$1-\sqrt{2x-1} \geq 0 \text{ if } x \leq 1, \text{ and } 1-\sqrt{2x-1}<0 \text{ if } x>1.$$

Consequently, if $\frac{1}{2} \leq x \leq 1$ then

(2) $$B = \frac{1}{\sqrt{2}}\left(1+\sqrt{2x-1}+1-\sqrt{2x-1}\right)=\sqrt{2},$$

and for $x>1$

(3) $$B = \frac{1}{\sqrt{2}}\left(1+\sqrt{2x-1}+\sqrt{2x-1}-1\right)=\sqrt{4x-2}.$$

Based on these observations

a) is satisfied if $\frac{1}{2} \leq x \leq 1$;

b) holds if $\sqrt{4x-2}=1$, i.e. $x=\frac{3}{4}$; consequently b) does not admit any solution;

c) holds if $\sqrt{4x-2}=2$, in other words $x=\frac{3}{2}$; this satisfies assumption (3) hence $x=\frac{3}{2}$ solves c).

1959/3. *Suppose that x satisfies*

(1) $$a\cos^2 x+b\cos x+c=0.$$

Form a quadratic equation in $\cos 2x$ whose roots are the same values of x. Apply your result for the special case $a=4$, $b=2$, $c=-1$.

Solution. Since $\cos^2 x = \dfrac{1+\cos 2x}{2}$, and so $\cos^4 x = \dfrac{1+2\cos 2x+\cos^2 2x}{4}$, by ordering and squaring (1) we can achieve that in the resulting expression only $\cos^4 x$ and $\cos^2 x$ appear:

$$b\cos x = -a\cos^2 x - c,$$

$$b^2 \cos^2 x = a^2 \cos^4 x + 2ac \cos^2 x + c^2.$$

Substitute $\cos^2 x$ and $\cos^4 x$:

$$a^2 \frac{1 + 2\cos 2x + \cos^2 2x}{4} + (2ac - b^2)\frac{1 + \cos 2x}{2} + c^2 = 0.$$

As a corollary we obtain

(2) $\qquad a^2 \cos^2 2x + (2a^2 + 4ac - 2b^2)\cos 2x + (a^2 + 4ac - 2b^2 + 4c^2) = 0;$

this is exactly the equation we were aiming for. With the given substitutions $a = 4, b = 2, c = -1$ we get:

$$4\cos^2 x + 2\cos x - 1 = 0,$$
$$16\cos^2 2x + 8\cos 2x - 4 = 0,$$

or in a simpler form:

(3) $\qquad 4\cos^2 2x + 2\cos 2x - 1 = 0.$

Remark. Notice that $\cos x$ and $\cos 2x$ are roots of the same quadratic equation. The reason for this is that from (3) we derive that $\cos x = \dfrac{-1 \pm \sqrt{5}}{4}$, hence $x = 72°$ or $144°$, consequently $2x = 144°$ and $288°$, but $\cos 288° = \cos(360° - 288°) = \cos 72°$.

1959/4. *Construct a right triangle with a given hypotenuse, if we know that the median corresponding to the hypotenuse is equal to the geometric mean of the two adjacent sides.*

Solution. According to the theorem of Thales, the median (corresponding to the hypotenuse) is equal to half of the hypotenuse, which is equal to the radius of the circumcircle of the triangle. Denote the two adjacent sides by a and b, the hypotenuse by $2R$ and the median by R. According to the assumptions we have

(1) $\qquad\qquad\qquad ab = R^2.$

The altitude corresponding to the hypotenuse will be denoted by m. Expressing twice the area of the right triangle in two different ways we get

(2) $\qquad\qquad\qquad R^2 = ab = 2R \cdot m,$

implying

$$m = \frac{R}{2}.$$

Now the construction goes as follows (see *Figure 1959/4.1*): considering the Thales circle over the hypotenuse $AB = 2R$, we intersect this with the line parallel to the hypotenuse of distance $\dfrac{R}{2}$. Any intersection of the circle and the line will provide a candidate for the vertex C (the four intersections obviously give

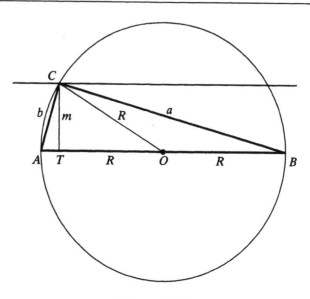

Figure 59/4.1

four congruent triangles). The triangle we got by this construction satisfies the assumption of the problem since (1) and (2) follow from $m = \dfrac{R}{2}$.

Remark. Notice that the angle $\angle COT$ in the right triangle OTC (see *Figure 1959/4.1*) is equal to $30°$, consequently the acute angles of ABC are equal to $15°$ and $75°$. Based on this observation other constructions are also possible.

1959/5. *A point M is moving on the interval AB. Squares $AMCD$ and $BMEF$ over the subintervals AM and MB (on the same side of the line AB) are constructed The circumcircles of these squares intersect each other in the points M and N. Verify that the lines AE and BC pass through N. Show that for any choice of M the line MN passes through a fixed point. Find the locus of the midpoints of the intervals joining the centres of the two squares.*

Solution. If M is the midpoint of AB then $C = E = N$, hence the first statement of the problem is obvious: $MN = f$ is the perpendicular bisector of AB, hence the common point of the lines MN (if such a point exists) should lie on f. Suppose now that $AM \neq MB$.

First we will show that C is the orthocentre of ABE (*Figure 1959/5.1*). In any case, EM is an altitude, and AC is orthogonal to BE since the diagonals of the two squares are parallel. Consequently C is the intersection of two altitudes, hence is the orthocentre. Now it follows that BC is also on the altitude, so BC is orthogonal to AE. Let N' denote the intersection of AE and BC. Since both diagonals AC and BE form right triangles with N', this point is on both the Thales circles corresponding to the diagonals, consequently N' is on the circumcircles of the squares. This, however, means that N' is equal to the common point of the two circles: $N' \equiv N$, hence AE and BC in fact pass through N.

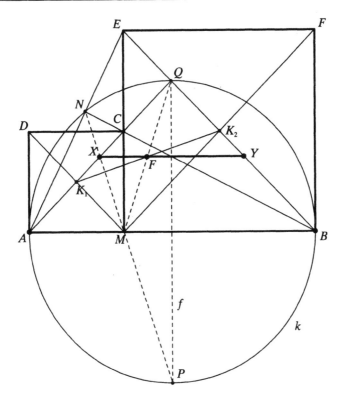

Figure 59/5.1

According to the famous connection between the angles at circumference and the central angles, the angle $\angle CNM$ is $45°$, hence NM is the bisector of the right angle of the triangle ANB. The circumcircle of ANB (denoted by k) coincides with the Thales circle of the interval AB, hence it is independent of the choice of M. Since the bisector at N intersects k in the midpoint P of the arc AB (not containing N), P is independent of the choice of M, hence all lines MN pass through P.

Let Q denote the point of intersection of AC and BE. Since Q is the third vertex of the isosceles right triangle ABQ constructed over AB, it is independent of the choice of M. If the centres of the squares are denoted by K_1 and K_2, then MK_2QK_1 is a rectangle. Hence the midpoint F of the diagonal K_1K_2 is also the midpoint of QM, so F is on the median XY of ABQ parallel to AB.

Consequently the locus we are looking for is the open interval XY: for arbitrary point F in XY there is a unique point M in AB (the point specified by twice the vector QF), and according to the above said the midpoint of the interval joining the centres of the squares determined by M is exactly F.

1959/6. *The planes P and Q intersect in the line p. A and C are points of the planes P and Q respectively, neither of them is on p. Construct that symmetric trapezium $ABCD$ (with $AB \parallel CD$) for which its vertices B and D are on the planes P and Q and $ABCD$ admits an incircle.*

Solution. Since AB and CD are parallel lines in two distinct, intersecting planes, they are necessarily parallel to p. The parallels to p through A in P (and through C in Q) will be denoted by e_1 and e_2 respectively; the construction will be performed in the plane determined by e_1 and e_2 (*Figure 1959/6.1*). Let d denote the distance of e_1 and e_2, and start with the trapezium constructed on *Figure 1959/6.2*. The orthogonal projection of C on e_1 will be denoted by C'. According to the theorem on the tangent intervals to a circle, $AC' = AD$ follows. Based on this information the trapezium can be constructed in the following manner:

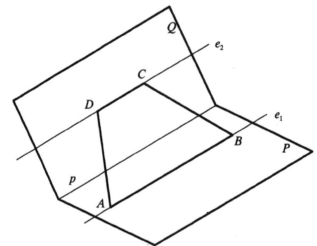

Figure 59/6.1

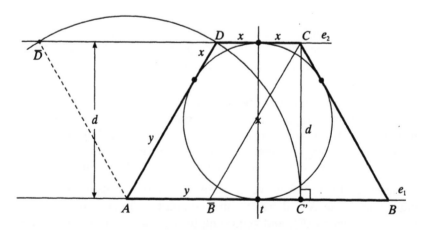

Figure 59/6.2

The circle of centre A and radius AC' intersects e_2 in D. Reflecting AD to the perpendicular bisector t of the interval CD we get BC. Now $ABCD$ satisfies the requirements, since (according to its construction) it is a symmetric trapezium, and since the sum of opposite sides are equal, it admits an incircle.

In case $AC' < d$, the problem has no solution. If $AC' = d$, there is a unique solution and in this case the trapezium turns out to be a square. Finally, if $AC' > d$, the circle with centre A and radius AC' intersects e_2 in two points, consequently in this case the problem has two solutions (see the trapezium $A\overline{B}C\overline{D}$ on the figure).

1960.

1960/1. *Determine all three digit numbers which are equal to 11 times the sum of the squares of their digits.*

Solution. Denote the digits by a, b and c ($a \neq 0$). According to the assumption

(1) $$100a + 10b + c = 11(a^2 + b^2 + c^2).$$

This implies that

$$(99a + 11b) + (a - b + c) = 11(a^2 + b^2 + c^2).$$

Now the right hand side and the first term on the left hand side is divisible by 11, hence so is $a - b + c$. Since $-18 \leq a - b + c \leq 18$, we conclude that $a - b + c$ is equal either to 0 or to 11.

In the first case $b = a + c$; substituting this expression into (1) we get
$$100a + 10(a + c) + c = 11(a^2 + (a + c)^2 + c^2).$$

After ordering, the quadratic equation

(2) $$2a^2 + (2c - 10)a + (2c^2 - c) = 0$$

follows. Since the first two terms of this expression are even, the third term should be even as well. This, however, implies that c is even. Equation (2) admits integer solutions if and only if its discriminant
$$4(-3c^2 - 8c + 25)$$
is a square. It is not hard to check that this is possible only in case $c = 0$. Substituting $c = 0$ in (2) we get $2a^2 - 10a = 0$, and since $a \neq 0$, we have $a = 5$. This implies
$$b = a + c = 5,$$
hence we get 550 for the integer we are seeking for; this number, in fact, satisfies the assumptions of the problem.

Let us now turn to the second case, when $b = a + c - 11$. After substituting and ordering (1) we get the expression

(3) $$2a^2 + (2c - 32)a + (2c^2 - 23c + 131) = 0.$$

This time it shows that c cannot be even. The discriminant admits the form
$$4(-3c^2 + 14c - 6);$$
for odd c it is a square only in case $c = 3$. Substituting this into (3),
$$2a^2 - 26a + 80 = 0$$
follows. By solving this quadratic equation we get that $a = 5$ or $a = 8$; based on (3) this implies $b = -3$ and $b = 0$ respectively. Obviously (since b is the middle

digit of the positive integer we are searching for), only the latter figure might serve as a solution of the original problem. This provides the second solution 803, and an easy check shows that this number satisfies the assumptions of the problem. In conclusion, the two solutions are 550 and 803.

1960/2. *Determine the real solutions of the inequality*

$$\frac{4x^2}{\left(1 \perp \sqrt{1+2x}\right)^2} < 2x+9.$$

Solution. The expression on the left hand side is defined if its denominator is nonzero, i.e., if $x \neq 0$; moreover if the expression under the square root is nonnegative, implying $1+2x \geq 0$, $x \geq \perp\frac{1}{2}$. Multiply both the numerator and the denominator of the fraction by $\left(1+\sqrt{1+2x}\right)^2$:

$$\frac{4x^2\left(1+\sqrt{1+2x}\right)^2}{4x^2} < 2x+9,$$

$$2\sqrt{1+2x} < 7,$$

$$4+8x < 49,$$

$$x < \frac{45}{8}.$$

Since all the manipulations above can be reversed, the solution set of the equation can be given by the inequalities

$$\perp\frac{1}{2} \leq x < 0, \qquad 0 < x < \frac{45}{8}.$$

1960/3. *The hypotenuse $BC = a$ of the right triangle ABC has been divided into n equal intervals with n an odd integer. Let h denote the altitude corresponding to the hypotenuse; and the central interval subtends an angle α at A. Show that*

$$\tan \alpha = \frac{4nh}{(n^2 \perp 1)a}.$$

First solution. We can assume that $n > 1$. The midpoint of BC will be denoted by F, while the two endpoints of the interval containing F are P and Q; let furthermore $AP = p$ and $AQ = q$. (See *Figure 1960/3.1.*) According to the theorem of Thales, $AF = \frac{a}{2}$. In the triangle APQ the interval AF is exactly the median corresponding to PQ and $PQ = \frac{a}{n}$, hence (see [1])

$$\frac{a^2}{4} = \frac{1}{4}\left(2p^2 + 2q^2 \perp \frac{a^2}{n^2}\right),$$

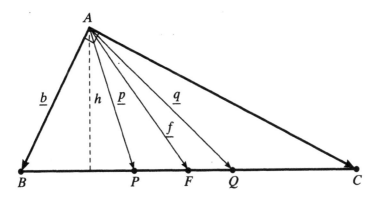

Figure 60/3.1

and this implies that

(1)
$$p^2 + q^2 = \frac{(n^2 + 1)a^2}{2n^2}.$$

Determine twice the area of APQ in two different ways and conclude

(2)
$$pq \sin \alpha = \frac{ah}{n}.$$

Apply the law of cosines for the triangle APQ; now (according to (1)) we get

(3) $$pq \cos \alpha = \frac{1}{2} \left(p^2 + q^2 \perp \frac{a^2}{n^2} \right) = \frac{1}{2} \left(\frac{(n^2 + 1)a^2}{2n^2} \perp \frac{a^2}{n^2} \right) = \frac{(n^2 \perp 1)a^2}{4n^2}.$$

Forming the ratio of appropriate sides of equations (2) and (3) we receive the expression

$$\tan \alpha = \frac{pq \sin \alpha}{pq \cos \alpha} = \frac{ah \cdot 4n^2}{n(n^2 \perp 1)a^2} = \frac{4nh}{(n^2 \perp 1)a},$$

which concludes the proof.

Second solution. A simpler solution can be given by using vector calculus. The vectors pointing form A to the points of the hypotenuse will be denoted by lower case bold face letters corresponding to the label of their endpoints. Since **f** corresponds to the midpoint of the hypotenuse, we know that

$$\mathbf{f} = \frac{\mathbf{b} + \mathbf{c}}{2},$$

consequently

$$\mathbf{p} = \mathbf{f} + \overrightarrow{FP} = \frac{\mathbf{b} + \mathbf{c}}{2} + \frac{\mathbf{b} \perp \mathbf{c}}{2n} = \frac{(n+1)\mathbf{b} + (n \perp 1)\mathbf{c}}{2n},$$

and similarly

$$q = f + \overleftrightarrow{FQ} = \frac{b+c}{2} + \frac{c \perp b}{2n} = \frac{(n \perp 1)b + (n+1)c}{2n}.$$

Since **b** and **c** are orthogonal, the above observation provides

$$pq = \frac{(n^2 \perp 1)(b^2 + c^2)}{4n^2} = \frac{(n^2 \perp 1)a^2}{4n^2}$$

which leads to

$$\tan \alpha = \frac{pq \sin \alpha}{pq \cos \alpha} = \frac{pq \sin \alpha}{pq} = \frac{ah \cdot 4n^2}{n(n^2 \perp 1)a^2} = \frac{4nh}{(n^2 \perp 1)a^2}$$

(by applying (2)). The proof is now complete.

1960/4. *Construct the triangle ABC if its two altitudes m_a and m_b (corresponding to the vertices A and B) and the median corresponding to A is given.*

Solution. Suppose that the triangle is constructed. Denote the feet of m_a and m_b by P and R. A' stands for the midpoint of BC and the foot of the orthogonal to AC from A' is Q (*Figure 1960/4.1*). Notice that $A'Q = \frac{m_b}{2}$ (since it is the line joining midpoints in the triangle BRC); furthermore the angles APA' and AQA' are both right angles, consequently P and Q are on the Thales circle over AA'.

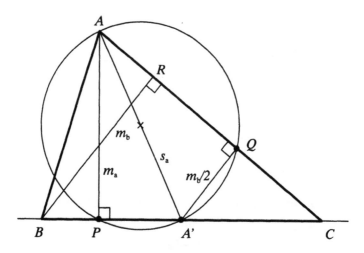

Figure 60/4.1

These observations already dictate the method of construction: construct the circle k with diameter $AA' = s_a$; now the circles with radius $\frac{m_b}{2}$ around A' and with radius m_a around A intersect k in Q and P, respectively. The point of intersection of the lines AQ and $A'P$ is C, while the reflection of C to A' gives B. The resulting triangle ABC contains AA' as its median; $AP = m_a$ is its altitude (corresponding to A) and the altitude corresponding to B is equal to $2 \cdot \dfrac{m_b}{2} = m_b$. Consequently, we verified that the resulting triangle solves the problem.

The construction can be carried out once the intersections exist, i.e., if $m_a \leq$ $\leq s_a$, $\dfrac{m_b}{2} \leq s_a$ and AQ is not parallel to $A'P$. In general, there are two solutions since the circles we used in the construction intersect the circle k in two points.

In case $m_a = s_a > \dfrac{m_b}{2}$ the points A' and P coincide, $A'P$ is tangent to the circle and the solution is a unique isosceles triangle.

If $m_a < s_a = \dfrac{m_b}{2}$ then $Q = A$, now AQ becomes tangent.

If $m_a = s_a = \dfrac{m_b}{2}$, there is no solution since the two tangents are parallel.

Finally, if $s_a > \dfrac{m_b}{2} = m_a$, there is a unique solution since for one choice of intersections the lines AQ and $A'P$ are parallel.

1960/5. *For a given cube $ABCDA'B'C'D'$ let X be a point of the face diagonal AC, and Y a point of $B'D'$.*

a) *Find the locus of the midpoints of the intervals XY for all possible choices of X and Y.*

b) *Consider the point $Z \in XY$ satisfying the equality $ZY = 2XZ$ and determine the locus of these points for all choices of X and Y.*

Solution. In solving the problem we will use the following result:

Suppose that α, β and γ are three parallel planes in the three-space, and γ is between α and β; suppose furthermore that it divides the distance between α and β as $a:b$. If X and Y are on the planes α and β respectively, then the locus of those points Z which divide XY as $a:b$ (more precisely, those which satisfy $XZ:ZY = a:b$) coincides with γ (see *Figure 1960/5.1*).

Our first aim is to find the locus of those points which divide XY (as given in the problem) according to the ratio $a:b$. The two faces $ABCD$ and $A'B'C'D'$ of the given cube are in parallel planes, hence (according to the result

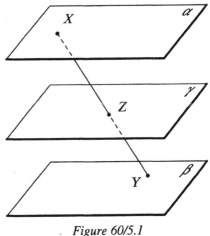

Figure 60/5.1

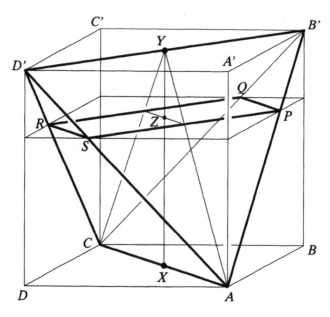

Figure 60/5.2

quoted above) the points Z we are looking for are elements of a plane γ parallel to these faces and dividing their distance as $a : b$ (*Figure 1960/5.2*).

A, C, B' and D' can be regarded as the vertices of a tetrahedron inscribed in our cube. Since a tetrahedron is a convex solid, any point of an interval joining its two points is contained by its interior or boundary. Consequently the intervals XY and so the points Z are in the interior or on the boundary of this tetrahedron. In conclusion, the locus we are searching for is a subset of the intersection of the plane γ and the tetrahedron described above.

It is easy to see that this intersection is a rectangle. This follows from the fact that for two intersecting planes the intersection of them with a third plane parallel to their intersection line is a union of two parallel lines; for the planes $AB'D'$ and $CB'D'$ these are parallel to $B'D'$, and for $B'AC$ and $D'AC$ to AC. So the opposite sides of the intersection are parallel; the adjacent ones are orthogonal since the diagonals $B'D'$ and AC are parallel to them and they happen to be orthogonal to each other.

In the figure the four vertices of this rectangle are denoted by P, Q, R, S. Next we will show that every interior and boundary point of this rectangle belongs to the locus we are searching for. For this, consider Z in the rectangle. A plane is determined by the line parallel to AC passing through Z together with the line AC. The intersection of this plane with the tetrahedron defines a triangle; its vertex Y opposite to AC is on the diagonal $B'D'$. The line YZ is contained by the plane determined by the triangle ACY, therefore it intersects

AC in a point X. Since Z is in γ, according to the result quoted above we have $XZ:ZY = a:b$, hence Z belongs to the locus we are looking for.

In summary, the locus asked by the problem is a rectangular section of the tetrahedron $AB'CD'$ with a plane parallel to the faces $ABCD$ and $A'B'C'D'$ dividing their distance as $a:b$.

In part a) of the problem $a = b$, hence γ is the mid parallel plane, the vertices of the resulting rectangle are the midpoints of the edges AB', $B'C$, CD' and $D'A$. In part b) the vertices of the rectangle are equal to the points dividing the above listed edges as $1:2$.

We just note here that in part a) the resulting rectangle is actually a square, since its edges are equal to half of the diagonal of the cube, and the four vertices of the square are the centres of four faces of the cube.

1960/6. *A cone of revolution has an inscribed sphere tangent to the base of the cone (and to the sloping surface of the cone). A cylinder is circumscribed about the sphere so that its base lies in the base of the cone. Let V_1 and V_2 denote the volume of the cone and the resulting cylinder, respectively.*

a) *Show that V_1 is not equal to V_2.*

b) *Determine the smallest possible value of $k = \dfrac{V_1}{V_2}$, and for this minimal k construct the half angle of the symmetric cone.*

First solution. Consider a plane containing the symmetry axes of the cone, the sphere and the cylinder. Take the intersection of the configuration with this plane. The resulting configuration consists of an isosceles triangle, its incircle and a square circumscribed about the circle (*Figure 1960/6.1*). The base of the triangle is equal to the diagonal $2R$ of the base circle of the cone, while the radius of its incircle equals the radius r of the sphere.

Recall that the radius of the incircle in a triangle is equal to the ratio of the area and half of the circumference of the given

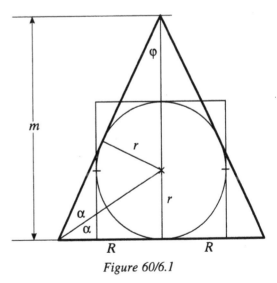

Figure 60/6.1

triangle. Denoting the altitude of the triangle (i.e., the cone) by m we get

$$r = \frac{Rm}{R + \sqrt{R^2 + m^2}}.$$

This implies $r\sqrt{R^2 + m^2} = R(m - r)$, hence

$$R^2 = \frac{r^2 m}{m - 2r}.$$

Since $V_1 = \dfrac{r^2 m^2 \pi}{3(m - 2r)}$ and $V_2 = 2r^3 \pi$,

$$k = \frac{V_1}{V_2} = \frac{m^2}{6(mr - 2r^2)} = \frac{1}{6\left(\frac{r}{m} - 2\frac{r^2}{m^2}\right)};$$

now the value of k is minimal if the denominator is maximal. Working with the denominator we have

$$-12\left(\left(\frac{r}{m} - \frac{1}{4}\right)^2 - \frac{1}{16}\right) = -12\left(\frac{r}{m} - \frac{1}{4}\right)^2 + \frac{3}{4};$$

we conclude that its maximum is $\dfrac{3}{4}$, hence the minimum of k is $\dfrac{4}{3}$ and it is attained when $\dfrac{r}{m} = \dfrac{1}{4}$, i.e., when $m = 4r$.

Based on these observations, the angle of the cone can be constructed by the following way: construct the two tangents to a circle of radius r from a point which is of distance $3r$ from the centre of the circle.

Notice that the above result already contains the solution of part a): since the minimum value of k is $\dfrac{4}{3}$, it never takes the value 1, so $V_1 = V_2$ is not possible.

Second solution. Using the notations of *Figure 1960/6.1* and denoting the angle of the apothems and the base plane by 2α we get

$$r = R\tan\alpha, \qquad m = R\tan 2\alpha,$$

hence the volumes of the cone and the cylinder are

$$V_1 = \frac{R^3 \pi \cdot \tan 2\alpha}{3}, \qquad V_2 = 2\pi R^3 \tan^3 \alpha,$$

respectively. This implies that

$$k = \frac{V_1}{V_2} = \frac{\tan 2\alpha}{6\tan^3 \alpha} = \frac{2\tan\alpha}{6\tan^3 \alpha(1 - \tan^2 \alpha)} =$$

$$= \frac{1}{3(-\tan^4 \alpha + \tan^2 \alpha)} = \frac{1}{-3\left(\tan^2 \alpha - \frac{1}{2}\right)^2 + \frac{3}{4}}.$$

Having k expressed in the above form we immediately see that $k \geq \dfrac{4}{3}$ (since the maximum of the denominator is $\dfrac{3}{4}$). This fact already implies that $k = 1$, i.e., $V_1 = V_2$ is not possible.

For k minimal we have $\tan^2 \alpha = \dfrac{1}{2}$, and since α is an acute angle, $\tan \alpha = \dfrac{\sqrt{2}}{2}$ and $\tan 2\alpha = \dfrac{2\tan\alpha}{1 \perp \tan^2 \alpha} = 2\sqrt{2}$. This implies

$$\tan \varphi = \tan(90° \perp 2\alpha) = \cot 2\alpha = \frac{1}{2\sqrt{2}},$$

and φ can be easily to constructed from this identity.

1960/7. *The parallel sides of a symmetric trapezium are of length a and b, while its altitude is m.*

a) *Construct the point P on the symmetry axis of the trapezium which is on the Thales circles over the legs of the trapezium.*

b) *Determine the distance of P from one of the parallel sides.*

c) *Under what assumption does such P exist?*

Solution. Let us denote the trapezium by $ABCD$; its base $AB = a$ is intersected by the symmetry axis in Q, while $CD = b$ intersects the axis in S. For a point P on the axis the angles $\angle APD$ and $\angle BPC$ are right angles if P is on the Thales circles over the edges AD and BC (*Figure 1960/7.1*); with this observation at hand the construction of P is straightforward.

Let $PQ = x$, then $PS = m \perp x$. CSP and PQB are similar right angled tri-

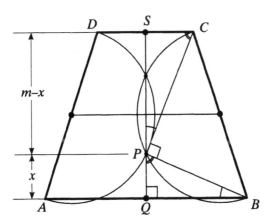

Figure 60/7.1

angles (since their acute angles are angles with orthogonal sides). This similarity implies that

$$(m \perp x): \frac{b}{2} = \frac{a}{2} : x,$$

from which it follows that

$$x^2 \perp mx + \frac{ab}{4} = 0,$$

$$x = \frac{m}{2} \pm \sqrt{\frac{m^2 \perp ab}{4}}.$$

The resulting two values of x are equal to the two distances of P from the parallel sides. (Notice that their sum is exactly m.) P exists if the expression under the square root is nonnegative, i.e., $m^2 \geq ab$. In case $m^2 > ab$, there are two points satisfying the requirements of the problem. These two points (as the construction already shows) are symmetric to the line joining the midpoints of the legs.

If $m^2 = ab$ then $x = \frac{m}{2}$, hence the Thales circle is tangent to the axis of symmetry; it is fairly easy to see that in this case the trapezium admits an incircle.

1961.

1961/1. *Solve the following system of equations for x, y and z:*

(1) $$x + y + z = a,$$
(2) $$x^2 + y^2 + z^2 = b^2,$$
(3) $$xy = z^2,$$

where a and b are given real numbers. What conditions must a and b satisfy for x, y and z to be all positive and distinct?

Solution. Subtract (2) from the square of (1) and (according to (3)) substitute xy by z^2:

$$2xy + 2xz + 2yz = a^2 \perp b^2,$$
$$2z^2 + 2z(x + y) = a^2 \perp b^2.$$

Now substitute $x + y$ with $a \perp z$ (as shown by (1)) and get

$$2z^2 + 2az \perp 2z^2 = a^2 + b^2.$$

$a = 0$ implies $b = 0$, hence from (2) we get $x = y = z = 0$, which — according to our additional hypothesis — is not a solution. Consequently we may assume that

$a \neq 0$, so

$$z = \frac{a^2 - b^2}{2a}, \qquad z^2 = \frac{(a^2 - b^2)^2}{4a^2}.$$

Now $x + y$ and xy can be expressed in terms of a and b as follows:

(4) $$x + y = a - z = \frac{a^2 + b^2}{2a} \quad \text{and} \quad xy = z^2 = \frac{(a^2 - b^2)^2}{4a^2}.$$

This yields a quadratic equation in u:

$$4a^2 u^2 - 2a(a^2 + b^2)u + (a^2 - b^2)^2 = 0,$$

which is solved by x and y. The solutions are

(5)
$$u_{1,2} = \frac{1}{8a^2}\left(2a(a^2 + b^2) \pm \sqrt{4a^2(a^2 + b^2)^2 - 16a^2(a^2 - b^2)^2} \right) =$$

$$= \frac{1}{4a}\left(a^2 + b^2 \pm \sqrt{(3a^2 - b^2)(3b^2 - a^2)} \right).$$

Hence the solutions of the system (under the hypothesis $a \neq 0$) are:

$$x = u_1, \qquad y = u_2, \qquad z = \frac{a^2 - b^2}{2a},$$

$$x = u_2, \qquad y = u_1, \qquad z = \frac{a^2 - b^2}{2a},$$

and these expressions obviously satisfy the equations.

Let us examine what are the necessary and sufficient conditions for the solutions to exist. (1) implies $a > 0$; for $z > 0$ we also need $a > |b|$. (4) implies that if the solutions exist then both x and y are positive. x and y are distinct if the discriminant in (5) is positive; since $a > |b|$, this positivity implies $|b|\sqrt{3} > a$. In this case neither x nor y is equal to z: if, for example, $x = z$ then (according to (3)) we have $y = z$ so $x = y$, which is impossible. In summary, the system admits solutions with the desired properties if

$$|b| < a < |b|\sqrt{3}.$$

Remark. Let us sketch the geometric background of the problem: (1) is simply the equation of a plane intersecting the axes in points of distance a from the origin, consequently its distance from the origin is $\dfrac{a}{\sqrt{3}}$. (2) is the equation of a sphere of radius $|b|$ around the origin; the two objects have common points if and only if $|b| \geq \dfrac{a}{\sqrt{3}}$, i.e., $a \leq |b|\sqrt{3}$. (3) is the equation of a cone with the origin as its vertex.

1961/2. *Let a, b and c be the sides of a given triangle while t is its area. Show that*

(1) $$a^2 + b^2 + c^2 \geq 4t\sqrt{3}$$

When does equality hold?

First solution. Suppose that the largest angle of the triangle ABC is at C. The foot of the altitude m at C is T, which is an inner point of the interval AB. Let x denote AT (*Figure 1961/2.1*). Apply the area-formula $cm = 2t$ and express a^2 and b^2 using the Pythagorean theorem:

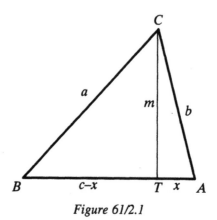

Figure 61/2.1

$$a^2+b^2+c^2-4t\sqrt{3}=(m^2+(c-x)^2)+(m^2+x^2)+c^2-2\sqrt{3}cm=$$
$$=2c^2+2m^2+2x^2-2cx-2\sqrt{3}cm=\frac{1}{2}\left[(c-2x)^2+(c\sqrt{3}-2m)^2\right]\geq 0,$$

providing the solution of the problem.

Equality holds only if $x=\dfrac{c}{2}$ and $m=\dfrac{c\sqrt{3}}{2}$, meaning that the triangle is equilateral.

Second solution. (1) is equivalent to
$$(a^2+b^2+c^2)^2\geq 48t^2=3\cdot 16t^2.$$

To demonstrate this, apply an appropriate version of the formula of Heron (see [16]):
$$16t^2=-a^4-b^4-c^4+2a^2b^2+2b^2c^2+2c^2a^2.$$

With this observation we only need to verify that
$$a^4+b^4+c^4+2a^2b^2+2b^2c^2+2c^2a^2\geq -3a^4-3b^4-3c^4+6a^2b^2+6b^2c^2+6c^2a^2,$$

which is equivalent to the inequality
$$(a^2-b^2)^2+(b^2-c^2)^2+(c^2-a^2)^2\geq 0.$$

This latter inequality, however, obviously holds and equality holds if and only if $a=b=c$.

Third solution. Let us invoke the formula of Heron again. From the inequality between the arithmetic and geometric means we have

$$(s \perp a)(s \perp b)(s \perp c) \le \left(\frac{(s \perp a)+(s \perp b)+(s \perp c)}{3}\right)^3 = \frac{s^3}{27},$$

hence

(2) $$t = \sqrt{s(s \perp a)(s \perp b)(s \perp c)} \le \sqrt{\frac{s^4}{27}} = \frac{s^2}{3\sqrt{3}},$$

and equality holds only for equilateral triangles. This last inequality implies

$$4t\sqrt{3} \le 3\left(\frac{2s}{3}\right)^2 = 3\left(\frac{a+b+c}{3}\right)^2 \le 3\frac{a^2+b^2+c^2}{3} = a^2+b^2+c^2,$$

and again equality holds only in case $a=b=c$.

Fourth solution. Apply the area formula $2t = ab \sin \gamma$ and $c^2 = a^2 + b^2 \perp 2ab \cos \gamma$, which is a direct consequence of the law of cosines:

$$a^2+b^2+c^2 \perp 4t\sqrt{3} = 2a^2 + 2b^2 \perp 4ab\left(\frac{\sqrt{3}}{2}\sin\gamma + \frac{1}{2}\cos\gamma\right) =$$

$$= 2a^2 + 2b^2 \perp 4ab\sin(\gamma+30°) \ge$$
$$\ge 2a^2 + 2b^2 \perp 4ab = 2(a \perp b)^2 \ge 0.$$

Equality holds if $a = b$, $\gamma = 60°$, i.e., when the triangle is equilateral.

Fifth solution. Our inequality can be substantially refined; for this we use the following identity valid for positive reals x, y, z:

(3) $$\sqrt{3xyz(x+y+z)} \le xy + yz + zx.$$

Now start with the obvious inequality

$$0 \le (xy \perp yz)^2 + (yz \perp zx)^2 + (zx \perp xy)^2.$$

After squaring and ordering we get

$$3x^2yz + 3xy^2z + 3xyz^2 \le x^2y^2 + y^2z^2 + z^2x^2 + 2x^2yz + 2xy^2z + 2xyz^2,$$
$$3xyz(x+y+z) \le (xy+yz+zx)^2,$$

which already implies the statement we aimed to prove. Equality holds when $x = y = z$.

Apply now (3) with $x = s \perp a$, $y = s \perp b$, $z = s \perp c$, $x+y+z = s$ and simplify with the formula of Heron expressing $t\sqrt{3}$:

$$t\sqrt{3} = \sqrt{3(s \perp a)(s \perp b)(s \perp c)s} \le (s \perp a)(s \perp b)+(s \perp b)(s \perp c)+(s \perp c)(s \perp a)$$

$$= 3s^2 \perp s(2a+2b+2c) + ab + bc + ca =$$

$$= \perp s^2 + ab + bc + ca = \frac{1}{4}(\perp(a+b+c)^2 + 4ab + 4bc + 4ca) =$$

$$= \frac{1}{4}(a^2+b^2+c^2 \perp (a \perp b)^2 \perp (b \perp c)^2 \perp (c \perp a)^2).$$

This gives

(4) $$4t\sqrt{3} \leq a^2 + b^2 + c^2 \perp (a \perp b)^2 \perp (b \perp c)^2 \perp (c \perp a)^2,$$

with equality if $a = b = c$. Now (1) obviously follows from (4).

Sixth solution. This last solution reveals the origin of the problem and

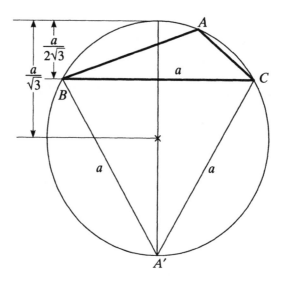

Figure 61/2.2

shows that it is obviously true for those triangles which admit an angle not less than $120°$. Let the angle of the triangle ABC at A not less than $120°$ and let $BC = a$. Construct the equilateral triangle $A'BC$ with sides a. The radius of the circumcircle of this triangle is $\dfrac{a}{\sqrt{3}}$ (*Figure 1961/2.2*). Put the triangle ABC to the side of BC opposite to A'. Then A is on or inside the circumcircle of the equilateral triangle, hence the altitude corresponding to A is at most half of the radius, so the area of ABC satisfies

$$t \leq \frac{1}{2}a \cdot \frac{a}{2\sqrt{3}} = \frac{a^2}{4\sqrt{3}}.$$

This implies $4t\sqrt{3} \leq a^2$, so $4t\sqrt{3} \leq a^2 + b^2 + c^2$ is valid for ABC.

In case the angles of ABC are all less than $120°$, then it contains a point I in its interior (the so-called isogonal point, see [4]) from which the angles $\angle AIB$, $\angle AIC$ and $\angle BIC$ are all equal to $120°$ (*Figure 1961/2.3*). According to the previous case now the triangles IAB, IBC and ICA have areas not more than $\dfrac{c^2}{4\sqrt{3}}$, $\dfrac{a^2}{4\sqrt{3}}$ and $\dfrac{b^2}{4\sqrt{3}}$, respectively, hence for the area of ABC we have

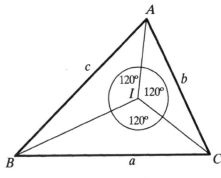

$$t \le \frac{a^2}{4\sqrt{3}} + \frac{b^2}{4\sqrt{3}} + \frac{c^2}{4\sqrt{3}}$$

implying (1). Equality can hold if the small triangles are isosceles, i.e., when I is the centre of the circumcircle, meaning that ABC is equilateral.

Figure 61/2.3

Remarks. Because of the fundamental importance of (1) in trigonometry it has relations with many other well-known inequalities. Using those, many more alternative solutions would be possible to give, here we merely list a few relevant connections. (1) was published by R. Weitzenböck in 1919, the refinement in (4) is due to P. Finsler and H. Hadwiger (1937).

1. By constructing equilateral triangles over the sides a, b and c inside the given triangle, the centres of those form an equilateral triangle with side square

$$\frac{1}{6}(a^2 + b^2 + c^2 \perp 4t\sqrt{3}).$$

2. For the sum of the squares of the sides we have ([2]):

$$a^2 + b^2 + c^2 = 4t(\cot\alpha + \cot\beta + \cot\gamma).$$

This implies that (1) is equivalent to the cotangent inequality ([3])

$$\cot\alpha + \cot\beta + \cot\gamma \ge \sqrt{3}.$$

3. An inequality due to F. Goldner states

$$4t\sqrt{3} \le \sqrt{3(a^2b^2 + b^2c^2 + c^2a^2)} \le a^2 + b^2 + c^2.$$

4. If r, r_a, r_b and r_c are the radii of the incircle and the excircles then (4) is equivalent to

$$r(r_a + r_b + r_c) \ge t\sqrt{3}.$$

5. Equation (2) in the third solution means that among triangles with a fixed circumference $2s$ the equilateral triangle admits maximal area; the value of this maximum is $\dfrac{s^2}{3\sqrt{3}} = \dfrac{s^2}{3}\cot\dfrac{\pi}{3}$. The generalization of this fact also holds: among

n-gons of fixed circumference $2s$ the regular n-gon has maximal area, and the value of this maximum is equal to $\dfrac{s^2}{n}\cot\dfrac{\pi}{n}$. Based on this, the method of the third solution shows that if the sides of an n-gon are equal to $a_1, a_2, \ldots, a_n$ and its area is t then

$$a_1^2 + a_2^2 + \ldots + a_n^2 \geq 4t\cot\frac{\pi}{n}.$$

1961/3. *Solve the equation*

$$\cos^n x \perp \sin^n x = 1$$

where n is a positive integer.

Solution. Suppose first that $n=1$. In this case our equality is

$$\cos x \perp \sin x = \sqrt{2}\left(\frac{1}{\sqrt{2}}\cos x \perp \frac{1}{\sqrt{2}}\sin x\right) = \sqrt{2}\cos\left(\frac{\pi}{4}+x\right) = 1,$$

$$\cos\left(\frac{\pi}{4}+x\right) = \frac{\sqrt{2}}{2},$$

from which $x = 2k\pi$ and $x = (4k+3)\dfrac{\pi}{2}$ follows. (Here and in the following k stands for an arbitrary integer.)

Let us consider the case $n \geq 2$ now. Rewrite the equation as

$$\cos^n x \perp \sin^n x = \cos^2 x + \sin^2 x,$$
$$\sin^2 x(1 + \sin^{n-2} x) + \cos^2 x(1 \perp \cos^{n-2} x) = 0.$$

Factors in the products on the left hand side are all nonnegative, hence the terms are nonnegative. Their sum can be equal to zero only if they are equal to zero:

(1) $\qquad\qquad\qquad \sin^2 x(1 + \sin^{n-2} x) = 0.$

(2) $\qquad\qquad\qquad \cos^2 x(1 \perp \cos^{n-2}) = 0.$

In case (1) there are two possibilities:

         a) $\sin x = 0,$ b) $\sin^{n-2} x = \perp 1.$

a) In this case $x = k\pi$. In (2) then we have $\cos^2 x = 1$, consequently (2) holds if and only if $\cos^{n-2} x = 1$. For n even this always holds, for n odd, however, only if $\cos x = 1$, i.e., $x = 2k\pi$.

b) This can happen only for n odd, and then $\sin x = \perp 1$, i.e., $x = (4k+3)\dfrac{\pi}{2}$. For these x we have $\cos x = 0$, hence (2) always holds.

In summary we got

    in case n even: $x = k\pi,$

    in case n odd: $x = 2k\pi$ and $x = (4k+3)\dfrac{\pi}{2}.$

1961/4. *P is a point inside the triangle $P_1 P_2 P_3$. The intersections of the lines $P_1 P$, $P_2 P$ and $P_3 P$ with the opposite sides are denoted by Q_1, Q_2 and*

Q_3, respectively. Show that there is one among the ratios

$$\frac{P_1P}{PQ_1},\ \frac{P_2P}{PQ_2},\ \frac{P_3P}{PQ_3}$$

which is not less, and one which is not more than 2.

First solution. Consider the parallel to P_2P_3 passing through the centre of gravity (denoted by S) of the triangle. This line intersects P_1P_2 in X_1 and P_1P_3

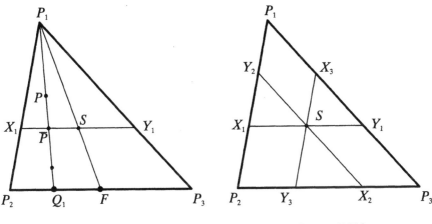

Figure 61/4.1   *Figure 61/4.2* .

in Y_1 (*Figure 1961/4.1*). Since X_1Y_1 divides the median P_1F as $2:1$, it divides all intervals joining P_1 with an arbitrary point of P_2P_3 in the same manner. X_1Y_1 divides the triangle into a triangle $P_1X_1Y_1$ and a trapezium $X_1Y_1P_3P_2$; we will call these domains the triangle- and trapezium-domain corresponding to P_1.

If P is in the triangle-domain then $\frac{P_1P}{PQ_1}\le 2$, and if it is in the trapezium-domain then $\frac{P_1P}{PQ_1}\ge 2$. Similar consideration holds for the two other vertices P_2 and P_3.

In order to prove the statement, we only need to show that for an arbitrary choice of P it belongs to one of the triangle- and one of the trapezium-domains. By studying *Figure 1961/4.2* it is easy to verify that $P_1P_2P_3$ is covered by both the triangle- and the trapezium-domains, and this observation concludes our proof.

Second solution. Following the lines of the first solution given above, we show that P is contained by both a triangle- and a trapezium-domain. It is easy to see that P is contained by the triangle-domain corresponding to P_1 if and only if $\frac{PQ_1}{P_1Q_1}\ge\frac{1}{3}$ and it is in the trapezium-domain of P_1 if and only if $\frac{PQ_1}{P_1Q_1}\le\frac{1}{3}$

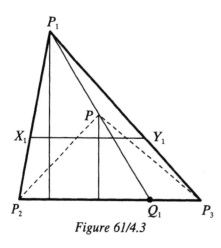

(*Figure 1961/4.3*). Notice, however, that the ratio of PQ_1 and P_1Q_1 is equal to the ratio of the altitudes of the triangles PP_2P_3 and $P_1P_2P_3$ corresponding to P_2P_3. Since the two triangles share their base, this latter ratio is equal to the ratio of their area. Let t_1, t_2 and t_3 denote the area of the triangles PP_2P_3, PP_3P_1 and PP_1P_2, respectively. For the area t of $P_1P_2P_3$ we obviously have $t_1 + t_2 + t_3 = t$, hence

$$\frac{PQ_1}{P_1Q_1} = \frac{t_1}{t}, \quad \frac{PQ_2}{P_2Q_2} = \frac{t_2}{t}, \quad \frac{PQ_3}{P_3Q_3} = \frac{t_3}{t}.$$

Since

$$\frac{t_1}{t} + \frac{t_2}{t} + \frac{t_3}{t} = \frac{t}{t} = 1,$$

among the three ratios there is one which is not more and one which is not less than $\frac{1}{3}$, concluding the solution.

Figure 61/4.3

1961/5. *Construct a triangle ABC if the length of the two sides $AC = b$ and $AB = c$ and the acute angle $AMB = \omega$ is given — here M is the midpoint of BC. Show also that the problem admits a solution if and only if*

$$b \tan \frac{\omega}{2} \le c < b.$$

Solution. Let us start with the triangle already constructed (*Figure 1961/5.1*). The interval $AB = c$ subtends an angle ω at the point M. Let us de-

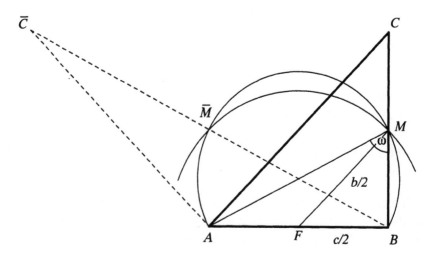

Figure 61/5.1

note the midpoint of AB by F. The length of FM is equal to $\frac{b}{2}$, hence M is on the circle of radius $\frac{b}{2}$ with centre F.

Now the method of the construction is as follows: Consider the circle k of points such that the interval $AB = c$ subtends an angle ω at them and intersect k with the circle s of radius $\frac{b}{2}$ centered at F. Their point of intersection is exactly M. By measuring the length of BM twice on the line BM passed M we get C. The resulting triangle ABC obviously satisfies the assumptions since AM is a median, hence $AC = 2FM = b$ and $\angle AMB = \omega$.

The condition of solvability is exactly that the circles k and s do intersect. Since ω is an acute angle, the Thales circle over AB is entirely inside k. Consequently the circle s intersects k only if its radius is at least $\frac{c}{2}$, i.e., $b > c$. For the existence of the intersection we also need that the farthest point

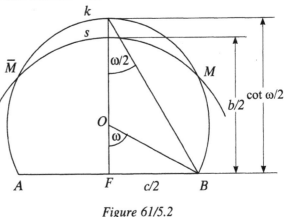

Figure 61/5.2

of s from AB is not farther away than the farthest point of k from AB. These distances from AB are $\frac{b}{2}$ and $\frac{c}{2} \cot \frac{\omega}{2}$ (*Figure 1961/5.2*), consequently the condition for solvability is

$$\frac{b}{2} \le \frac{c}{2} \cot \frac{\omega}{2}, \quad \text{hence} \quad b \tan \frac{\omega}{2} \le c.$$

In conclusion, the necessary and sufficient condition we are looking for is

$$b > c \ge b \tan \frac{\omega}{2}.$$

Once this assumption is satisfied, we have two solutions (in *Figure 1961/5.1* the second solution is indicated by dashed lines).

1961/6. *Let ε be a given plane and A, B, C three non-collinear points on one side of ε. Suppose furthermore that the plane determined by these points is parallel to ε. Let A', B' and C' be three arbitrary points on ε. The midpoints of AA', BB' and CC' are denoted by L, M and N respectively. The centre of gravity of the triangle LMN is denoted by G. (Those triples A', B', C' for which L, M, N do not form a triangle, are disregarded.) Determine the locus of G for any possible choice of the triple A', B', C' on the plane ε.*

Solution. Consider the centre of gravity of ABC as origin (*Figure 1961/6.1*). A vector pointing from O to a point denoted by an upper case letter will be labeled by the same lower case letter. From the definition of the centre of mass it follows that

$$\mathbf{a}+\mathbf{b}+\mathbf{c}=\mathbf{0}.$$

The given data imply that

$$\mathbf{l}=\frac{\mathbf{a}+\mathbf{a}'}{2}, \qquad \mathbf{m}=\frac{\mathbf{b}+\mathbf{b}'}{2}, \qquad \mathbf{n}=\frac{\mathbf{c}+\mathbf{c}'}{2},$$

hence the vector pointing to the centre of gravity of LMN is

$$\mathbf{g}=\frac{1}{3}\left(\frac{\mathbf{a}+\mathbf{a}'}{2}+\frac{\mathbf{b}+\mathbf{b}'}{2}+\frac{\mathbf{c}+\mathbf{c}'}{2}\right)=\frac{1}{6}\left((\mathbf{a}+\mathbf{b}+\mathbf{c})+(\mathbf{a}'+\mathbf{b}'+\mathbf{c}')\right)=\frac{1}{2}\left(\frac{\mathbf{a}'+\mathbf{b}'+\mathbf{c}'}{3}\right).$$

Let S' denote the centre of gravity of the triple A', B', C'. (The notion of centre of gravity is extended for collinear points as follows: the centre of gravity of A', B', C' is the endpoint of the vector $\dfrac{\mathbf{a}'+\mathbf{b}'+\mathbf{c}'}{3}$.) According to the above said $\mathbf{g}=\dfrac{1}{2}\mathbf{s}'$, hence G is in the plane ε' we get by applying the enlargement with ratio $\dfrac{1}{2}$ centered at the origin. In other words, ε' is the plane which is parallel to ε and divides the distance of O and ε by two.

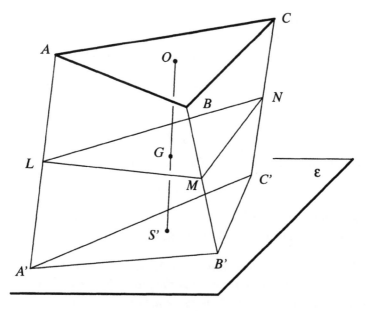

Figure 61/6.1

Next we have to show that any point of ε' can be got in this way. Let G be an arbitrary point of ε' and denote the intersection of OG and ε by S'. Choose

$A' \in \varepsilon$ in such a way that AA' does not pass through G, hence the midpoint L of AA' is not equal to G. Fix a line passing through B disjoint both from LG and $A'S'$, intersecting ε in B'. The midpoint of BB' is denoted by M. Choose C' in such a way that S' becomes the centre of gravity of the triangle defined by A', B' and C'. Finally, let N denote the midpoint of CC'. According to the above said, the vector pointing to the centre of gravity of LMN is half of s', consequently it points to G. L, M and N are not collinear — otherwise G would be on the line containing the three points, but this fact contradicts the choice of L and M.

1962.

1962/1. *Determine the smallest possible positive integer x whose last decimal digit is 6, and if we erase this last 6 and put it in front of the remaining digits, we get four times x.*

First solution. Let the digits of the originally n-digit number x denoted by $a_{n-1}, a_{n-2}, \ldots, a_1, a_0 = 6$, so

$$x = 10^{n-1}a_{n-1} + 10^{n-2}a_{n-2} + \ldots + 10a_1 + 6$$

According to the problem

$$4x = 6 \cdot 10^{n-1} + \left(10^{n-2}a_{n-1} + 10^{n-3}a_{n-2} + \ldots + 10a_2 + a_1\right)$$

The expression in the parenthesis clearly equals to $\dfrac{x \perp 6}{10}$, therefore

$$4x = 6 \cdot 10^{n-1} + \frac{x \perp 6}{10}.$$

This implies

$$40x = 6 \cdot 10^n + x \perp 6,$$
$$13x = 2(10^n \perp 1).$$

This shows that $10^n \perp 1$ (a number containing only 9) must be divisible by 13; by inspecting these numbers we see that $999\,999$ is the first which is divisible by 13:

$$999\,999 = 13 \cdot 76\,923,$$

therefore

$$x = 2 \cdot 76\,923 = 153\,846,$$

and this x solves the problem.

Second solution. It is easy to see that if the last digit of a number is 6, then four times this number has 4 as its last digit, so if x is of the form $\ldots 6$

then $4x$ admits the form $\ldots 4$. According to the problem $4x$ has been created by erasing the last digit of x, consequently x must be of the form $\ldots 46$. Now the last two digits of $4x$ is easy to determine: $4x = \ldots 84$. Using this principle we can determine more and more digits of both x and $4x$:

x	$4x$
$\ldots .46$	$\ldots .84$
$\ldots 846$	$\ldots 384$
$\ldots 3\,846$	$\ldots 5\,384$
$.53\,846$	$.15\,384$
$153\,846$	$615\,384$

This scheme now shows that $153\,846$ satisfies the assumptions of the problem, and it is the smallest such number.

1962/2. *Determine all real x satisfying*

$$\sqrt{3-x}-\sqrt{x+1}>\frac{1}{2}.$$

Solution. By ordering the inequality we get

$$\sqrt{3-x}>\frac{1}{2}+\sqrt{x+1}.$$

The expressions appearing here are defined only in the interval $-1\leq x\leq 3$; moreover in this interval all terms in the inequality are positive. Therefore the above inequality is equivalent to its square

$$3-x>\frac{1}{4}+x+1+\sqrt{x+1},$$

$$\sqrt{x+1}<\frac{7}{4}-2x.$$

Since the left hand side is nonnegative, the right hand side is necessarily positive; for this reason

(1) $$-1\leq x<\frac{7}{8}$$

has to be satisfied. For such x squaring the inequality is a reversible transformation, hence

$$x+1<\frac{49}{16}-7x+4x^2,$$

$$(x-1)^2>\frac{31}{64},$$

$$|x-1|>\frac{\sqrt{31}}{8},$$

from which

$$x > 1 + \frac{\sqrt{31}}{8} \quad \text{or} \quad x < 1 \perp \frac{\sqrt{31}}{8}$$

follows. Only the latter one obeys assumption (1); since we only performed equivalent transformations, the solution set of our inequality is equal to

$$\perp 1 \le x < 1 \perp \frac{\sqrt{31}}{8}.$$

We just note here that because of $1 \perp \frac{\sqrt{31}}{8} < \frac{7}{8}$, the inequality $x < \frac{7}{8}$ is automatically satisfied.

1962/3. *The cube $ABCDA'B'C'D'$ with upper face $ABCD$ and lower face $A'B'C'D'$ $(AA' \parallel BB' \parallel CC' \parallel DD')$ is given. A point X runs along the perimeter of $ABCD$ (in the direction given by the above order) with constant speed, while a point Y does the same (with equal speed) along the perimeter of the square $B'C'CB$. X and Y start in the same instant from A and B', respectively. Determine the locus of the midpoint Z of the interval XY.*

Solution. We will use the following fact: for AB and CD arbitrary unit intervals with X (and Y) moving on them from A to B (from C to D, respectively) with equal speed, the midpoint F of XY sweeps the line segment joining the midpoints of the segments AC and BD.

In order to verify this statement, let $\mathbf{a}$ denote the vector pointing to A while $\mathbf{c}$ points to C. If $\overrightarrow{AB} = \mathbf{e}_1$ and $\overrightarrow{CD} = \mathbf{e}_2$ *(Figure 1962/3.1)* then the position of X and Y is determined by the vectors

$$\mathbf{x} = \mathbf{a} + \lambda \mathbf{e}_1, \qquad \mathbf{y} = \mathbf{c} + \lambda \mathbf{e}_2 \qquad (0 \le \lambda \le 1),$$

hence the vector $\mathbf{f}$ pointing to F is equal to

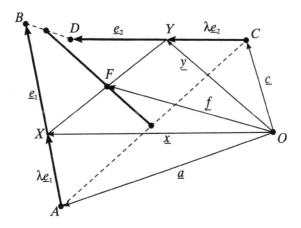

Figure 62/3.1

(1)
$$f = \frac{a+c}{2} + \lambda \frac{e_1 + e_2}{2}.$$

Consequently, F is on the interval with starting point the midpoint of AC ($\lambda = 0$) and direction $\frac{e_1 + e_2}{2}$. The endpoint of this interval is exactly the midpoint of BD, since in case $\lambda = 1$ we have

$$f = \frac{(a+e_1) + (c+e_2)}{2},$$

and this vector points to the midpoint of BD. Each point of this interval is the midpoint of an interval XY, since for each f (as given in (1)) one can find the appropriate x and y.

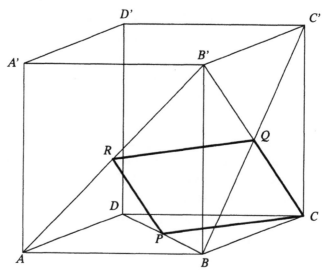

Figure 62/3.2

Now the motion described by the problem can be subdivided into four portions, each corresponding to the case of a pair of intervals. These individual cases are taken care of by our argument above (*Figure 1962/3.2*). Call the midpoint of BD, BC' and AB' by P, Q and R respectively. The four portions of the motion of X and Y can be listed as follows:

	the interval run by X	the interval run by Y
1.	AB,	$B'C'$,
2.	BC,	$C'C$,
3.	CD,	CB,
4.	DA,	BB'.

According to our starting argument, the interval swept by the midpoint is RQ in the first case, QC in the second, CP in the third and finally PR in the fourth case. All these points belong to the locus we are looking for. In summary, the locus is equal to the quadrilateral $PCQR$. By considering the orthogonal projection to the face $ABCD$ one can show that $PCQR$ is a parallelogram; since both QC and CP are equal to the half of the diagonal of this face, the parallelogram is in fact a rhombus. Since to diagonals of adjacent faces starting from the same vertex of the cube determine an angle of $60°$, the acute angle of our rhombus is $60°$ as well.

1962/4. *Solve the following equation:*
$$\cos^2 x + \cos^2 2x + \cos^2 3x = 1.$$

Solution. Use the following substitutions:
$$\cos 2x = 2\cos^2 x \perp 1, \qquad \cos 3x = \cos x(4\cos^2 x \perp 3),$$
and order the equation:
$$\cos^2 x + (2\cos^2 x \perp 1)^2 + \cos^2 x(4\cos^2 x \perp 3)^2 = 1,$$
$$\cos^2 x(8\cos^4 x \perp 10\cos^2 x + 3) = 0,$$
$$\cos^2 x(2\cos^2 x \perp 1)(4\cos^2 x \perp 3) = 0,$$
$$\cos x \cos 2x \cos 3x = 0.$$

In conclusion, the solution set consists of those x for which $\cos x$, $\cos 2x$ or $\cos 3x$ vanishes. Since $\cos x = 0$ implies $\cos 3x = 0$, it is sufficient to determine the solutions of $\cos 2x = 0$ and $\cos 3x = 0$.
$$\cos 2x = 0, \quad \text{if} \quad 2x = (2k+1)\frac{\pi}{2}, \quad x = (2k+1)\frac{\pi}{4},$$
$$\cos 3x = 0, \quad \text{if} \quad 3x = (2k+1)\frac{\pi}{2}, \quad x = (2k+1)\frac{\pi}{6},$$

where k is an arbitrary integer. Since all the transformations we have performed are reversible, the set of values determined above is the solution of the problem.

1962/5. *Three distinct points A, B and C on a circle k are given. Construct the point D on the circle for which the quadrilateral ABCD admits an incircle.*

First solution. Recall that a quadrilateral admits an incircle if and only if the sum of opposite edges are equal.

Suppose that D is a point of the circumcircle of the triangle ABC which is on the arc AC not containing B, and for which $AB + CD = BC + AD$ (i.e., $ABCD$ admits an incircle, see *Figure 1962/5.1*). If $AB = BC$ then $CD = AD$ also holds consequently D is the midpoint of the arc AC, and since $ABCD$ is a kite, it admits an incircle.

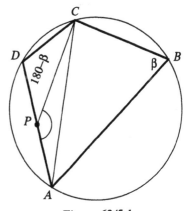

Figure 62/5.1

Without loss of generality we may assume that $AB > BC$, hence $AB \perp \perp BC = AD \perp CD > 0$. Measure CD from D onto DA and call its endpoint P. The triangle CDP is isosceles with the third angle being equal to $180° \perp \beta$, hence the angles on its base are equal to $\frac{\beta}{2}$, and so the angle APC is $180° \perp \frac{\beta}{2}$.

Based on the above said, the construction of D proceeds as follows: construct the circle such that AC subtends angle $180° \perp \frac{\beta}{2}$, and consider the arc opposite to B. Intersect this arc with the circle of radius $AB \perp BC$ and centre A, and call the point of intersection P. By construction $AP = AB \perp BC$. Now the line determined by AP intersects the circumcircle of ABC in the point D we are seeking for.

Next we show that the quadrilateral $ABCD$ admits an incircle. Since $\angle CDA = 180° \perp \beta$ and $\angle CPA = 180° \perp \frac{\beta}{2}$, the angle $\angle CPD$ is equal to $\frac{\beta}{2}$, hence $\angle PCD = \frac{\beta}{2}$. Consequently CDP is an isosceles triangle, so $CD = DP$, implying that $AD = AP + CD = AB \perp BC + CD$. This shows that $AB + CD = AD + BC$, which means that $ABCD$ admits an incircle.

The existence of the construction depends on the existence of P. This point, however, always exists, since the circle such that AC subtends an $180° \perp \frac{\beta}{2}$ is inside the circumcircle; moreover the circle with radius $AP = AB \perp BC$ and centre A always intersects the inner arc since $AP = AB \perp BC < AC$ according to the triangle inequality applied to ABC. Consequently the problem admits a unique solution.

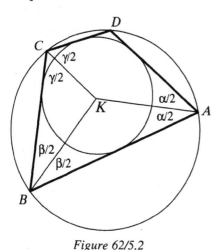

Figure 62/5.2

Second solution. In this solution we utilize the fact the if a quadrilateral admits an incircle then the bisectors intersect each other in a unique point K (the centre of the incircle), see *Figure 1962/5.2*.

Suppose that the quadrilateral has been constructed and the angles of it at the vertices A, B and C are α, β and γ, respectively. $\alpha + \gamma = 180°$ since $ABCD$ admits a circumcircle; because $\angle AKB = 180° \perp \frac{\alpha+\beta}{2}$ and $\angle BKC = 180° \perp \frac{\beta+\gamma}{2}$, we have

$$\angle AKC = 360° \perp \left(180° \perp \frac{\alpha+\beta}{2}\right) \perp \left(180° \perp \frac{\beta+\gamma}{2}\right) = \frac{2\beta+\alpha+\gamma}{2} = 90° + \beta.$$

If $\beta = 90°$ then ABC is a right angled triangle and D is the reflection of B to its hypotenuse AC.

If $\beta < 90°$ then the centre point K is the intersection of the angle bisector at B with the circle such that AC subtends angle $90° + \beta$. Having constructed K, the incircle of the quadrilateral can be easily constructed since it is tangent to BA and BC. Now the tangents of the incircle passing through A and C intersect each other in D.

The resulting quadrilateral obviously admits an incircle — we need to show that it also admits a circumcircle. According to the construction the sum of the three angles at K is

$$360° = 90° + \beta + \left(180° \perp \frac{\beta+\gamma}{2}\right) + \left(180° \perp \frac{\beta+\alpha}{2}\right) = 450° \perp \frac{\alpha+\gamma}{2},$$

implying

$$\alpha + \gamma = 180°,$$

consequently $ABCD$ admits a circumcircle, i.e., D is on the circumcircle of ABC.

Finally, if $\beta > 90°$ the construction has to be modified only in one point: we need to consider the circle such that AC subtends angle $270° \perp \beta$. It is straightforward that in all cases the construction yields a unique solution.

Remark. The quadrilateral under consideration is a so-called bicentric quadrilateral since it admits both an incircle and a circumcircle. Such quadrilaterals have many interesting properties, for example, two data among the radii of the incircle, the circumcircle and the distance of their centre determine the third one (see [5]).

1962/6. *Let R and r denote the radii of the circumcircle and the incircle of an isosceles triangle. Show that the distance d between the centres of the two circles is*

(1) $$d = \sqrt{R(R \perp 2r)}.$$

First solution. Let ABC be an isosceles triangle ($AC = BC$), and let O be the centre of its circumcircle while K is the centre of its incircle. The bisector at C intersects the circumcircle for the second time in the midpoint Q of the arc AB; the foot of the orthogonal from K to AC is T (*Figure 1962/6.1*).

We start with the observation that $QA = KQ$. This can be shown by demonstrating that AKQ is an isosceles triangle: $\angle AKQ = \frac{\alpha}{2} + \frac{\gamma}{2}$ and $\angle QAB = \angle QCB = \frac{\gamma}{2}$, hence $\angle QAK = \frac{\alpha}{2} + \frac{\gamma}{2}$, verifying that AKQ is an isosceles triangle.

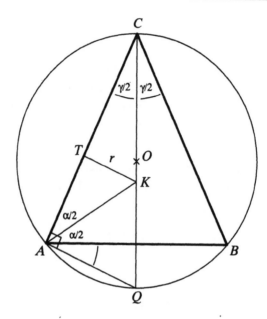

Figure 62/6.1

Let $OK = d$. Based on the formula regarding the power of K to the circumcircle we have (see [6])

$$(2) \qquad KQ \cdot KC = R^2 \perp d^2.$$

Using the similarity of the right angled triangles CAQ and CTK we get

$$(3) \qquad \frac{r}{QA} = \frac{KC}{CQ} = \frac{KC}{2R},$$

implying

$$2Rr = QA \cdot KC,$$

and since $QA = KQ$, this shows

$$2Rr = KQ \cdot KC.$$

Comparing it with (2) we conclude that

$$2Rr = R^2 \perp d^2,$$
$$d = \sqrt{R(R \perp 2r)}.$$

Second solution. We claim that (1) holds for any triangle. We prove it by vector methods.

Let us fix the origin at the centre of the circumcircle; now the vectors pointing to the vertices of the triangle are **a**, **b** and **c**. By definition we have $|\mathbf{a}| = |\mathbf{b}| = |\mathbf{c}| = R$. It is known that the vector pointing to the centre of the incircle is equal to (see [7])

$$\mathbf{k} = \frac{a\mathbf{a} + b\mathbf{b} + c\mathbf{c}}{a + b + c} = \frac{a\mathbf{a} + b\mathbf{b} + c\mathbf{c}}{2s},$$

where a, b and c stand for the lengths of the sides of the triangle. According to our notation we have that $|\mathbf{k}| = d$. Consider the square (with inner product!) of the equality

$$2s\mathbf{k} = a\mathbf{a} + b\mathbf{b} + c\mathbf{c}.$$

This leads to

(2) $$4s^2 d^2 = (a^2 R^2 + b^2 R^2 + c^2 R^2 + 2ab\mathbf{ab} + 2bc\mathbf{bc} + 2ca\mathbf{ca}),$$

now applying of $|\mathbf{a} \perp \mathbf{b}| = c$, i.e., $(\mathbf{a} \perp \mathbf{b})^2 = c^2$ yields

$$2R^2 \perp 2\mathbf{ab} = c^2,$$

which is equivalent to $2\mathbf{ab} = 2R^2 \perp c^2$. Similarly, we have $2\mathbf{bc} = 2R^2 \perp a^2$ and $2\mathbf{ca} = 2R^2 \perp b^2$. Substitute these into (2) and conclude:

$$4s^2 d^2 = R^2 (a+b+c)^2 \perp abc(a+b+c) = R^2 \cdot 4s^2 \perp abc \cdot 2s,$$

giving

$$d^2 = R^2 \perp \frac{abc}{2s}.$$

According to the radius formula (see [8]) $abc = 4tR$, $t = rs$, and so

$$d^2 = R^2 \perp 2Rr, \qquad d = \sqrt{R^2 \perp 2Rr},$$

concluding the solution.

Remarks. 1. Equation (1) was discovered by L. Euler; its important consequence is the so-called radius-inequality (see [37]):

$$R \geq 2r,$$

and equality holds if and only if $d = 0$, i.e., in case of an equilateral triangle.

2. (1) is in close connection to Poncelet's prisms (see [9]).

3. A slight modification of our first solution also gives that (1) holds for all triangles. If ABC is not an isosceles triangle, hence K and O are not on the same diameter of the circumcircle, we should consider the diameter QC', where Q is the midpoint of the arc AB; therefore the bisector CK passes through Q (*Figure 1962/6.2*). Now the proof should be modified at (3) by substituting CQ with $C'Q$ and in the previous line we should invoke the similarity of $C'AQ$ and CTK.

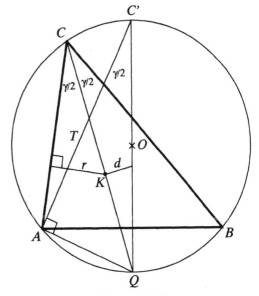

Figure 62/6.2

1962/7. *There are five spheres which are tangent to all extended edges of a tetrahedron $SABC$. Show that*

a) $SABC$ *is a regular tetrahedron;*

b) *conversely: a regular tetrahedron admits five spheres with the properties described above.*

Solution. If a sphere is tangent to the extended edges of a tetrahedron, then the faces intersect the sphere in circles which are tangent to the extended edges; consequently in plane sections the resulting circles are the incircles or excircles of the faces of the tetrahedron.

Two touching circles of a triangle cannot be on the same sphere and an excircle can share a sphere with an incircle only in case they touch the edge at the same point (*Figure 1962/7.1*).

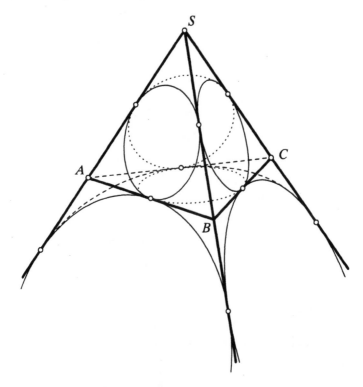

Figure 62/7.1

If we add the fact that two tangent circles lying in different planes uniquely determine the sphere containing them and that this sphere contains all circles tangent to these two circles (see [10]), we conclude that one of the five spheres (call it G) contains the four incircles of the faces, while the remaining four spheres contain one incircle and three excircles tangent to this circle. (Recall that two circles are said to be tangent if in their common point their tangents coincide.)

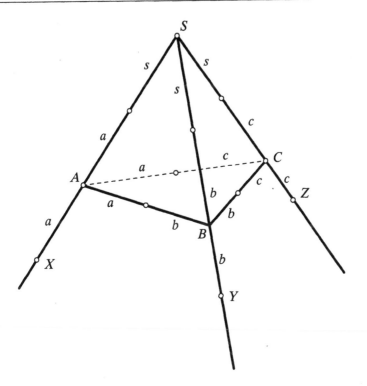

Figure 62/7.2

In order to show that $SABC$ is regular, it is enough to verify that any face of it is an equilateral triangle. Let us consider the face ABC. Mark the points on the edges of the tetrahedron where G intersects them. On edges starting from a common vertex these marked points are of equal distance from the vertex (since the tangents to a sphere from an external point are of equal length). In *Figure 1962/7.2* these equal intervals are denoted by the lower case letters corresponding to the labels of the vertices.

Let G_s be the sphere containing the incircle of ABC and the containing excircles of the other three faces. It intersects the extended edges SA, SB and SC in the points X, Y and Z, respectively. Since the length a of the tangent interval from A to the incircle of ABC is the same as the tangent interval from A to G_s, we have $AX = a$ and similarly $BY = b$, $CZ = c$. Using the above principle the tangent intervals from S to G_s are equal, consequently $SX = SY = SZ$, hence

$$s + 2a = s + 2b = s + 2c,$$

implying $a = b = c$. This shows that the sides of ABC are equal, hence it is an equilateral triangle.

Now if $SABC$ is a regular tetrahedron with centre K and the radius of the incircle is equal to r, then by denoting the radii of the incircles of the faces by ϱ we have that the distance of K from any edge is equal to $r' = \sqrt{r^2 + \varrho^2}$

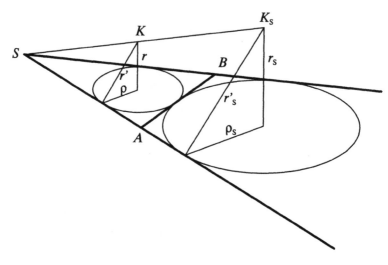

Figure 62/7.3

(*Figure 1962/7.3*). Therefore the sphere of radius r' and centre K is tangent to all extended edges.

Similarly, if the centre of the sphere touching the face ABC of the tetrahedron $SABC$ from outside is K_s and its radius is r_s, furthermore the radius of the excircle of ABC is ϱ_s, then the distance of K_s from the edges is equal to $r'_s = \sqrt{r_s^2 + \varrho_s^2}$. Consequently, the sphere with centre K_s and radius r'_s is tangent to all edges. This shows the existence of five spheres with the desired properties for a regular tetrahedron.

Remarks. 1. Studying *Figure 1962/7.2* a little longer, we might notice that if the tetrahedron admits a sphere which is tangent to the edges in their inner points then the sum of opposite edges are equal (on the figure this sum is $a + +b+c+s$). One can prove that this condition is sufficient for the existence of a sphere tangent to the edges.

2. Similar observations can be made regarding the existence of spheres which are external tangents of the edges: the necessary and sufficient condition for the existence of *one* such sphere is that the difference of any base edge and the facing edge should be constant. (Here base edges are the edges of the triangle which is tangent to the sphere from inside.) For tetrahedrons with equal faces this condition is fulfilled (see [17]), hence they admit four external tangent spheres, but the inner tangent exists only in case the tetrahedron is regular.

3. Our proof showed that the existence of the inner tangent sphere and the external tangent sphere tangent to the edges of the face ABC from inside imply that ABC is equilateral and SA, SB, SC are equal. In a similar vein it follows that if the external tangent sphere to the face SBC exists then $SA = AB = AC$. This however implies that all edges of the tetrahedron are equal, consequently it is regular. According to this observation the existence of the inner and two external tangent spheres already implies that $SABC$ is regular.

1963.

1963/1. *Determine the real solutions of the following equality (p denotes a real parameter)*

$$\sqrt{x^2 - p} + 2\sqrt{x^2 - 1} = x.$$

Solution. Since x is the sum of two nonnegative numbers, $x \geq 0$ obviously holds; moreover we need $x^2 \geq p$ and $x \geq 1$ for the square roots to make sense. Under these assumptions the equality

$$4\sqrt{x^2 - p}\sqrt{x^2 - 1} = p + 4 - 4x^2$$

is equivalent to our starting one.

The left hand side is nonnegative, hence $p + 4 - 4x^2 \geq 0$ has to be satisfied, implying $x^2 \leq \dfrac{p+4}{4}$.

With this assumption, another squaring yields

$$8(2 - p)x^2 = (p - 4)^2,$$

and this equation is equivalent to our starting one.

Now $p \neq 2$, since otherwise the left hand side vanishes while the right hand side does not. This shows that

(1) $$x^2 = \frac{(p-4)^2}{8(2-p)}.$$

Next we examine the possible values of p for which our previous (necessary) assumptions on x are satisfied.

a) $x^2 \geq 1$, i.e. $\dfrac{(p-4)^2}{8(2-p)} \geq 1$ or $\dfrac{p^2}{8(2-p)} \geq 0$, implying $p < 2$.

b) $x^2 \geq p$, i.e. $\dfrac{(p-4)^2}{8(2-p)} \geq p$ or $\dfrac{(3p-4)^2}{8(2-p)} \geq 0$, implying $p < 2$.

c) $x^2 \leq \dfrac{p+4}{4}$, i.e. $\dfrac{(p-4)^2}{8(2-p)} \leq \dfrac{p+4}{4}$ or $\dfrac{3p\left(p-\frac{4}{3}\right)}{8(2-p)} \leq 0$, implying $0 \leq p \leq \dfrac{4}{3}$.

The common part of these assumptions is

$$0 \leq p \leq \frac{4}{3}.$$

In conclusion, for these values of p the equation admits a unique solution of the form

$$x = \frac{4 - p}{\sqrt{16 - 8p}},$$

while for other values of p the equation has no solution at all.

1963/2. *Given a point A and a segment BC in the 3-dimensional space, determine the locus of those points, P, for which the angle ∠APX is a right angle for some X on the segment BC.*

Solution. Let X be an arbitrary point of BC. If one leg of the right angle contains A and the other one X, then the vertex P of the right angle is on the sphere with AX as its diameter, and all the points of this sphere satisfy the condition of the problem.

Once X runs through the points of BC, the corresponding spheres sweep out the locus M asked by the problem. $(*)$

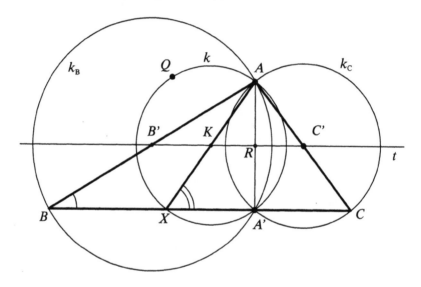

Figure 63/2.1

In the following we would like to give a better description of this set M. Let G_B and G_C denote the balls with diameters AB and AC, respectively. We would like to show that M is equal to the set of points which are not exterior to one and not interior to the other one.

Let us first examine a plane section of this set; let the plane S contain our three given points A, B and C. (Notice that if the three points are not collinear then S is determined by these points.) The centres of the spheres are on the interval joining the midpoint B' of AB with the midpoint C' of AC; the line $B'C' = t$ turns out to be a symmetry axis for each sphere in M, moreover by rotation around t any point of M might be transported into S. Let the circle k_B be the plane section of G_B with S while the intersection of S and G_C is denoted by k_C (*Figure 1963/2.1*).

In order to prove our statement, it is enough to verify that the main circles of the spheres in M fill up the domain we get by deleting the common interior points from the union of the discs k_B and k_C. In the following this domain will be denoted by T. The other intersection point of k_B and k_C will be denoted by A' (notice that it might coincide with A), in this notation t is the perpendicular bisector of AA'. Since any circle in S with centre on t which passes through A also passes through A', we conclude that the main circles under consideration all pass through A and A'.

It is easy to see that the points of k_B and k_C belong to M. Let now Q be an inner point of T, for example in the interior of k_B. The centre K of the circle k passing through A, A' and Q is on t, and the endpoint X of its diameter is on the interval BA'. This last assertion follows from the fact that the angle $\angle ABA'$ is less than the angle $\angle AXA'$, consequently Q is on a circle such that its diameter is an interval joining a point X of BC with A. This observation now shows that Q belongs to M. For similar reasons any circle with diameter AX belongs to T.

We just note here that (after slight modifications) our arguments are also valid in the case when A, B and C are collinear.

Remark. In fact, the solution of the problem could be terminated at the point indicated by (∗), since by that point we answered the question raised in the problem. Since a locus can be described in many different (equivalent) ways, the expectations for solutions of such problems always contain indeterminacy.

1963/3. *Consider a convex n-gon with equal angles and with consecutive sides $a_1, a_2, \ldots, a_n$ satisfying*

(1) $$a_1 \geq a_2 \geq \ldots \geq a_n.$$

Show that under the above conditions we have

(2) $$a_1 = a_2 = \ldots = a_n.$$

Solution. The equality of the angles imply that the sides of the n-gon are parallel to the sides of a regular n-gon. We have to prove that the n-gon itself is regular. Let $n = 2k$ or $n = 2k + 1$ and

$$a_n = b_1, a_{n-1} = b_2, \ldots, a_{n-k+1} = b_k, a_{n-k} = b_{k+1}.$$

Denoted the inner bisector at the common vertex of a_1 and $a_n = b_1$ by f (*Figure 1963/3.1*).

The endpoint of a_k coincides with one of the endpoints of b_k if n is even; for n odd the edge $a_{k+1} = b_{k+1}$ is orthogonal to f. For this reason the orthogonal projection of $a_1, a_2, \ldots, a_k$ to f coincides with the orthogonal projections of

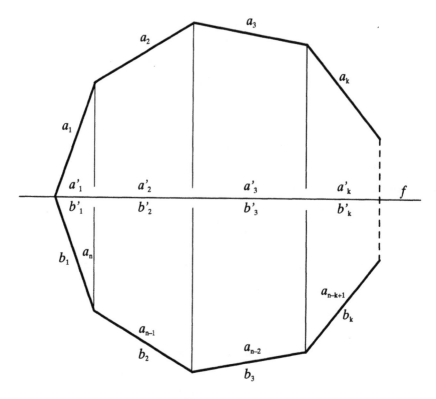

Figure 63/3.1

$b_1, b_2, \ldots, b_k$. The projections of a_i (b_i) will be denoted by a_i' (b_i'); so

(3) $$a_1' + a_2' + \ldots + a_k' = b_1' + b_2' + \ldots + b_k'.$$

Since the angle of a_i and b_i with f are equal, according to (1) we have

(4) $$a_i \geq b_i, \quad \text{hence} \quad a_i' \geq b_i'.$$

Now from (3) it follows that

$$(a_1' \perp b_1') + (a_2' \perp b_2') + \ldots + (a_k' \perp b_k') = 0;$$

since the terms in the parentheses are all nonnegative, their sum is zero only in case all terms are equal to zero. This shows $a_1' = b_1'$, implying $a_1 = b_1 = a_n$, consequently in (1) we have equality. This concludes our solution.

Remark. The statement of the problem can be visualized in the following way: consider a regular n-gon with side a_1. Starting from a_1 construct the n-gon given by the array (1); now if there is a place in (1) where $>$ is valid, then from that point our polygonal path is inside the regular n-gon, and will never close to form an n-gon.

1963/4. *Determine the values x_1, x_2, x_3, x_4, x_5 satisfying*

$$(1) \qquad x_5 + x_2 = yx_1,$$
$$(2) \qquad x_1 + x_3 = yx_2,$$
$$(3) \qquad x_2 + x_4 = yx_3,$$
$$(4) \qquad x_3 + x_5 = yx_4,$$
$$(5) \qquad x_4 + x_1 = yx_5$$

where y is a given parameter.

Solution. Express x_5 and x_3 from (1) and (2):

$$(6) \qquad x_5 = yx_1 - x_2,$$
$$(7) \qquad x_3 = yx_2 - x_1.$$

Substitute (6) into (5) and get

$$(8) \qquad x_4 = (y^2 - 1)x_1 - yx_2.$$

After substituting (8) and (7) into (3) and ordering the equation we get

$$(9) \qquad (y^2 + y - 1)(x_1 - x_2) = 0.$$

Substituting (6), (7) and (8) into (4) we see that

$$(10) \qquad (y^2 + y - 1)((y - 1)x_1 - x_2) = 0.$$

If $y^2 + y - 1 = 0$, i.e.

$$y = \frac{-1 \pm \sqrt{5}}{2},$$

then (9) and (10) is satisfied for arbitrary x_1 and x_2; since all our previous transformations are reversible, these values uniquely determine x_3, x_4, x_5.

If $y^2 + y - 1 \neq 0$ then (9) and (10) gives

$$(11) \qquad x_1 - x_2 = 0,$$
$$(y - 1)x_1 - x_2 = 0,$$

implying

$$(12) \qquad (y - 2)x_1 = 0.$$

For $y = 2$ the value of $x_1 = x_2$ can be arbitrary; if $x_1 = x_2 = c$ then $x_3 = x_4 = x_5 = c$.

Finally if $y \neq 2$ then $x_1 = 0$ from (12) and $x_2 = 0$ from (11) implying that all unknowns in the original system are equal to 0.

In summary:

1. If $y = (-1 \pm \sqrt{5})/2$ then $x_1 = a$, $x_2 = b$, $x_3 = yb - a$, $x_4 = (y^2 - 1)a - yb = -y(a + b)$, $x_5 = ya - b$, for a, b arbitrary reals.

2. If $y = 2$ then $x_1 = x_2 = x_3 = x_4 = x_5 = c$ for an arbitrary real c.

3. Finally if is y none of the above three values then $x_1 = x_2 = x_3 = x_4 = x_5 = 0$.

1963/5. *Show that*

(1)
$$\cos \frac{\pi}{7} \perp \cos \frac{2\pi}{7} + \cos \frac{3\pi}{7} = \frac{1}{2}.$$

First solution. Introduce $\alpha = \dfrac{\pi}{7}$ (so $7\alpha = \pi$). Consider the equilateral triangle OAB inside the angle α with vertex O and then consider the isosceles triangles ABC and BCD such that $OA = AB = BC = CD = 1$ is satisfied (see *Figure 1963/5.1*).

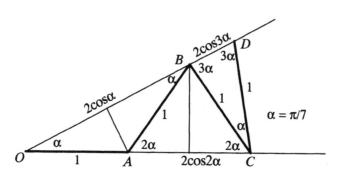

Figure 63/5.1

According to theorems regarding the sum of angles and external angles of a triangle, we conclude that the base angles of the above triangles are α, 2α and 3α, respectively. Moreover the angles of the triangle COD on CD are equal to 3α, hence it is an isosceles triangle, implying $OC = OD$. But

$$OC = 1 + 2\cos 2\alpha, \qquad OD = 2\cos \alpha + 2\cos 3\alpha,$$

hence

$$1 + 2\cos 2\alpha = 2\cos \alpha + 2\cos 3\alpha,$$

which obviously implies (1).

Second solution. By considering the identity

$$\cos \frac{5\pi}{7} = \cos \left(\pi \perp \frac{2\pi}{7} \right) = \perp \cos \frac{2\pi}{7}$$

(1) can be rewritten as

(2)
$$\cos \frac{\pi}{7} + \cos \frac{3\pi}{7} + \cos \frac{5\pi}{7} = \frac{1}{2}.$$

This identity, however, is connected to geometric properties of the regular heptagon. The symmetry axis of the regular heptagon with unit sides passing through A bisects the opposite side. The angle between the side and the perpendicular m to the axis in A is $\dfrac{\pi}{7} = \alpha$, hence the projection of this side on m is $\cos \alpha$ (*Figure*

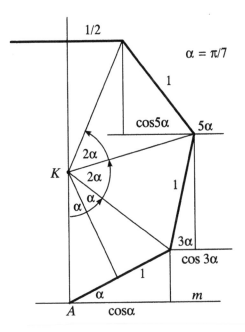

Figure 63/5.2

1963/5.2). Since consecutive sides can be got by rotating the previous side by 2α, the angles of the next two sides with m are 3α and 5α. Consequently the lengths of the projections are $\cos 3\alpha$ and $\cos 5\alpha$, this latter one being negative. The sum of these three oriented projections (signs are assigned according to the orientation) is equal to half of the side (see the figure), hence:

$$\cos \alpha + \cos 3\alpha + \cos 5\alpha = \frac{1}{2},$$

which is exactly what we wanted to demonstrate.

Third solution. The equality will be proved in the form given by (2). Denote the left hand side of (2) by K and apply

$$2 \sin x \cos y = \sin(x + y) \perp \sin(y \perp x).$$

Using the notation $\alpha = \dfrac{\pi}{7}$ and taking $\sin \alpha = \sin 6\alpha$ into account we get

$$K \cdot 2 \sin \alpha = 2 \sin \alpha \cos \alpha + 2 \sin \alpha \cos 3\alpha + 2 \sin \alpha \cos 5\alpha =$$
$$= \sin 2\alpha + \sin 4\alpha \perp \sin 2\alpha + \sin 6\alpha \perp \sin 4\alpha = \sin 6\alpha = \sin \alpha,$$

implying

$$K = \frac{1}{2}.$$

Remark. All three solutions above generalize to the proof of the following statement:

$$\text{If} \quad \alpha = \frac{\pi}{2k + 1} \quad (k = 1, \ 2, \ 3, \ \dots) \text{ then}$$

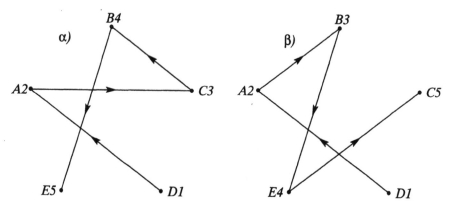

Figure 63/6.1c, d

$$\cos \alpha + \cos 3\alpha + \ldots + \cos(2k \perp 1)\alpha = \frac{1}{2}.$$

1963/6. *Five students, A, B, C, D and E were placed 1 to 5 in a contest. Someone made the initial guess that the final result would be the order ABCDE, but — as it turned out — this person was wrong on the final position of all the contestants; moreover no two students predicted to finish consecutively did so. A second person guessed DAECB, which was much better, since exactly two contestants finished in the place predicted, and two disjoint pairs predicted to finish consecutively did so. Determine the outcome of the contest.*

Solution. For simplicity, we assume that the competitors are vertices of a graph on five points, and two vertices are connected by an oriented edge if and only if the competitor corresponding to the starting vertex of the edge finished right before the student corresponding to the end vertex of the edge. Now any outcome can be represented by an oriented path containing four edges. We also label each vertex with its place.

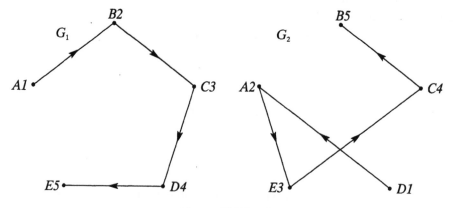

Figure 63/6.1a, b

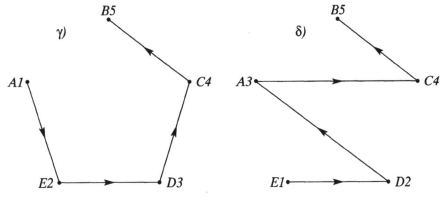

Figure 63/6.1e, f

$ABCDE$ corresponds to G_1 (*Figure 1963/6.1*) while $DAECB$ corresponds to G_2. We have to find a graph G which

a) does not share an edge or label with G_1 and

b) shares two edges and two labels with G_2.

According to our conventions, if G_2 and G share an edge and the labels on one of its endpoints are equal in the two graphs then the labels on the other endpoints are equal as well. Furthermore: G_2 and G cannot share two adjacent edges since in that case none of the three endpoints can have equal labels. Two adjacent edges cannot be common in G_2 and G even if the labels are different, because in that case the labels differ at least in four positions, which contradicts our assumption.

Notice that the common labels of G and G_2 are the endpoints of one of their common edges. Since among the remaining three vertices there is a common edge with different labels, the common edge connects vertices either with positions 1 and 2 or with 4 and 5. The second guess corresponds to the situation in which one of the two above edges is correct and the position of the remaining ones are cyclically permuted.

In conclusion the final result (based on the second guess) is among:

$$\alpha : DACBE, \qquad \beta : DABEC,$$
$$\gamma : AEDCB, \qquad \delta : EDACB.$$

In case α and γ the graphs G and G_1 share vertices with equal labels ($E5$ and $A1$), in β they share a common edge (AB). Consequently these do not satisfy the assumptions of the problem. δ however provides a solution, hence the outcome of the competition was $EDACB$.

1964.

1964/1. a) *Find all positive integers n for which 7 divides $2^n \mp 1$.*

b) *Show that there is no positive integer n for which 7 divides $2^n + 1$.*

Solution. After examining particular cases, one can see that divisibility of $2^n \pm 1$ by 7 is connected to divisibility of n by 3. Let $n = 3k + r$ $(r = 0, 1, 2)$; then

$$2^n = 2^{3k+r} = 2^r \cdot 8^k = 2^r(8^k \mp 1^k) + 2^r = 7E + 2^r,$$

where E stands for some integer, hence

$$2^n \mp 1 = 7E + 2^r \mp 1 \quad \text{and} \quad 2^n + 1 = 7E + 2^r + 1.$$

From this, if

$$
\begin{array}{lll}
r = 0, & 2^n \mp 1 = 7E, & 2^n + 1 = 7E + 1, \\
r = 1, & 2^n \mp 1 = 7E + 1, & 2^n + 1 = 7E + 3, \\
r = 2, & 2^n \mp 1 = 7E + 3, & 2^n + 1 = 7E + 5,
\end{array}
$$

This means that $2^n \mp 1$ is divisible by 7 if and only if n is divisible by 3 and $2^n + 1$ is not divisible by 7.

1964/2. *Let a, b and c denote the lengths of the sides of a triangle. Show that*

(1) $$a^2(\mp a + b + c) + b^2(a \mp b + c) + c^2(a + b \mp c) \leq 3abc.$$

First solution. Let us start with the following obvious inequality (the second factors, according to the triangle inequality, are all negative):

$$(b \mp c)^2(a \mp b \mp c) \leq 0,$$
$$(c \mp a)^2(b \mp c \mp a) \leq 0,$$
$$(a \mp b)^2(c \mp a \mp b) \leq 0.$$

Adding these inequalities we get:

$$a^2(b \mp c \mp a + c \mp a \mp b) + b^2(a \mp b \mp c + c \mp a \mp b) + c^2(a \mp b \mp c + b \mp c \mp a) \mp$$
$$\mp 2bc(a \mp b \mp c) \mp 2ca(b \mp c \mp a) \mp 2ab(c \mp a \mp b) =$$
$$= 2a^2(\mp a + b + c) + 2b^2(a \mp b + c) + 2c^2(a + b \mp c) \mp 6abc \leq 0,$$

from which (1) directly follows.

Second solution. After multiplication and ordering we get

$$\mp a^3 \mp b^3 \mp c^3 + a^2b + ab^2 + a^2c + ac^2 + b^2c + bc^2 \mp 2abc \leq abc;$$

factoring the left hand side we get

(2) $$(\mp a + b + c)(a \mp b + c)(a + b \mp c) \leq abc,$$

and this inequality is equivalent to the one we started with. Since in this last expression both sides are positive, it is equivalent to its square:

$$(\mp a + b + c)^2(a \mp b + c)^2(a + b \mp c)^2 \leq a^2 b^2 c^2,$$

$$(a^2 - (b - c)^2)(b^2 - (c - a)^2)(c^2 - (a - b)^2) \le a^2 b^2 c^2.$$

Now the terms on the left hand side are smaller (not larger) than the corresponding terms on the right hand side — verifying this last, hence our starting inequality.

Third solution. After ordering (1) divide both sides by $2abc$:

$$a(b^2 + c^2 - a^2) + b(a^2 + c^2 - b^2) + c(a^2 + b^2 - c^2) \le 3abc,$$

$$\frac{b^2 + c^2 - a^2}{2bc} + \frac{a^2 + c^2 - b^2}{2ac} + \frac{a^2 + b^2 - c^2}{2ab} \le \frac{3}{2}.$$

The quotients on the left hand side are equal to the cosines of the angles of the triangle, as it is shown by the law of cosines, hence the above inequality becomes

$$\cos \alpha + \cos \beta + \cos \gamma \le \frac{3}{2}.$$

This latter inequality is, however, the famous cosine inequality which is valid for all triangles, and (according to the above said) it is equivalent to (1) (see [12]).

Fourth solution. We show that (1) is satisfied by an arbitrary triple of nonnegative numbers a, b and c. Since their role in the inequality is symmetric, we can assume that

$$0 \le c \le b \le a.$$

In this case

$$a^2(b + c - a) + b^2(c + a - b) + c^2(a + b - c) - 3abc =$$
$$= a(ab + ac - a^2 - bc) + b(bc + ba - b^2 - ac) + c(ac + cb - c^2 - ab) =$$
$$= -a(a - b)(a - c) + b(b - c)(a - b) - c(a - c)(b - c) \le$$
$$\le -a(a - b)(b - c) + b(b - c)(a - b) - c(a - c)(b - c) =$$
$$= -(a - b)^2(b - c) - c(a - c)(b - c) \le 0.$$

Equality holds in case $a = b = c$ or $a = b$ and $c = 0$.

Remarks. 1. Solutions 1.–3. also show that equality holds only in case $a = b = c$, i.e., for an equilateral triangle.

2. The fact that (1) is equivalent to (2) — one of the most fundamental inequalities in trigonometry — indicates that our inequality has deep connections with numerous other geometric inequalities, as it was pointed out by the third solution. Another connection can be found by writing (2) in the form

$$8s(s - a)(s - b)(s - c) \le s \cdot abc,$$

and applying Heron's formula together with the identities $t = rs$, $abc = 4tR$ where r and R denote the radii of the incircle and the circumcircle and t is the area of the triangle. Now this rewriting yields

$$8t^2 \le s \cdot 4tR,$$
$$2r^2 s^2 \le s \cdot rsR,$$
$$2r \le R,$$

the famous radius inequality ([37]).

3. Adding $2(a^3 + b^3 + c^3)$ to the form of the inequality we found in the second solution we get

$$(a^2 + b^2 + c^2)(a + b + c) \leq 2(a^3 + b^3 + c^3) + 3abc,$$

still equivalent to (1). This form admits the following generalization: for a_i non-negative,

$$(a_1^{n-1} + a_2^{n-1} + \ldots + a_n^{n-1})(a_1 + a_2 + \ldots + a_n) \leq$$
$$\leq (n \perp 1)(a_1^2 + a_2^2 + \ldots + a_n^2) + na_1 a_2 \ldots a_n.$$

1964/3. *Let a, b, c denote the lengths of the sides of the triangle ABC. Tangents to the inscribed circle are constructed parallel to the sides. Each tangent forms a triangle with the other two sides of the triangle, and a circle is inscribed in each of these three triangles. Find the total area of all four inscribed circles.*

Consider the tangents of the incircle which are parallel to the sides. These tangents give rise to three subtriangles of ABC, consider the incircles of these subtriangles. Determine the sum of the areas of the four incircles.

Solution. Let the tangent of the incircle parallel to BC be denoted by $B'C'$ (where B' is on AB and C' is on AC). Let r_1 denote the radius of the incircle of $AB'C'$, while the radius and area of the incircle of ABC is denoted by r and t. Finally, m_a is the altitude of ABC corresponding to $BC = a$ is m_a (*Figure 1964/3.1*).

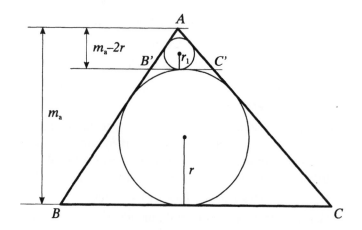

Figure 64/3.1

The triangles ABC and $AB'C'$ are similar, hence the ratio of the radii of the incircles coincides with the the ratio of the altitudes:

$$\frac{r_1}{r} = \frac{m_a \perp 2r}{m_a}, \qquad r_1 = r\left(1 \perp \frac{2r}{m_a}\right).$$

Since $2t = a m_a = 2rs$, $m_a = \dfrac{2rs}{a}$, this shows

$$(1) \qquad r_1 = r\left(1 \perp \frac{a}{s}\right) = \frac{r(s \perp a)}{s}.$$

Similar reasoning for the radii of the two other incircles provide $r_2 = \dfrac{r(s \perp b)}{s}$

and $r_3 = \dfrac{r(s \perp b)}{s}$. The sum of the four areas is equal to

$$T = \pi(r^2 + r_1^2 + r_2^2 + r_3^2) =$$

$$= \frac{\pi r^2}{s^2}\left(s^2 + (s \perp a)^2 + (s \perp b)^2 + (s \perp c)^2\right) = \frac{\pi r^2}{s^2}(a^2 + b^2 + c^2).$$

Since $r^2 = \dfrac{t^2}{s^2} = \dfrac{(s \perp a)(s \perp b)(s \perp c)}{s}$, we get

$$T = \frac{\pi(s \perp a)(s \perp b)(s \perp c)(a^2 + b^2 + c^2)}{s^3}.$$

Remark. An interesting corollary of (1) states that the sum of the radii of the incircles of the three subtriangles is equal to the radius of the incircle of the original triangle:

$$r_1 + r_2 + r_3 = r\left(1 \perp \frac{a}{s} + 1 \perp \frac{b}{s} + 1 \perp \frac{c}{s}\right) = r.$$

1964/4. *Each pair from* 17 *scientists exchange letters on one of three topics. Prove that there are at least three scientists who write to each other on the same topic.*

Solution. Consider a graph with 17 vertices and associate a vertex to each scientist. Two vertices are connected by an edge if the corresponding scientists exchange letters. The edge is red if they discuss the first topic, it is blue if the second, and it is green if they exchange letters on the third topic. We have to show that the graph contains a monochromatic triangle, i.e., there are three vertices connected by edges of the same colour.

First we will prove a useful lemma, sometimes also called Ramsey's Theorem (see our remark after the solution). The lemma says the following: Suppose that the edges of the complete graph on six vertices are coloured by two colours. Then the graph contains a monochromatic triangle.

Suppose that the colours we use are blue and green. Fix a vertex A. Among the five edges starting from A there are at least three of the same colour, say blue, joining A with B_1, B_2 and B_3. If one of the edges of the triangle $B_1 B_2 B_3$ is blue (for example $B_1 B_2$) then $A B_1 B_2$ is a monochromatic triangle. If none of the edges in $B_1 B_2 B_3$ is blue then all of them are green, hence we have found a monochromatic triangle again.

Let us now return to our original problem. Fix a vertex A of the graph on 17 vertices. There are at least 6 edges among the 16 edges starting from A which are decorated by the same colour, say red. Let the endpoints of these 6 edges be called B_1, B_2, B_3, B_4, B_5 and B_6. If one of the edges connecting these vertices (say $B_1 B_2$) is red, then $AB_1 B_2$ is monochromatic, so we are done.

If there are no red edges joining any two of $B_1, \ldots, B_6$ then all edges are blue or green, hence the lemma proved above shows that there is a monochromatic triangle among them. This last observation concludes the solution.

Remark. 1. The statement does not remain true for a graph with 16 vertices: such a complete graph admits a colouring with three colours such that no triangle is monochromatic.

2. The problem is one of the Ramsey-type theorems of graph theory; it admits many generalizations and applications.

3. The following generalization directly relates to our problem: let $a_1 = = 3$ and $a_n = na_{n-1} \perp (n \perp 2)$ for $n > 1$. If the complete graph on a_n vertices is coloured by n colours then it admits a monochromatic triangle.

For $n = 2$ this theorem is simply the lemma we used above, for $n = 3$ it is exactly the problem. The proof goes by induction: for $n = 2, 3$ the statement — as we verified above — is true. Suppose that it is true up to $n \perp 1$. Now consider the complete graph on a_n vertices coloured by n colours and fix a vertex A of it. The $a_n \perp 1 = na_{n-1} \perp n + 1$ edges starting from it are also coloured by n colours, so there are at least a_{n-1} which are coloured by the same colour.

Let the endpoints of the a_{n-1} vertices of the same colour (say red) starting from A be denoted by $B_1, B_2, \ldots B_{a_{n-1}}$. If there is a red edge among them (say between B_i and B_k), then $AB_i B_k$ is a red triangle, and we proved the assertion. If there is no red edge among $B_1, B_2, \ldots, B_{a_{n-1}}$, then it is a subgraph on a_{n-1} vertices coloured by $n \perp 1$ colours, so — according to the inductive hypothesis — it admits a monochromatic triangle.

It can be shown that

$$a_n = [n!e] + 1.$$

(See also the remark at problem 1978/6.)

1964/5. *Five points on the plane are situated so that no two of the lines joining a pair of points are coincident, parallel or perpendicular. Through each point lines are drawn perpendicular to each of the lines through two of the other four points. Give the best possible upper bound for the number of intersection points of these orthogonals, disregarding the given 5 points.*

Solution. Denote the five given points by A_1, A_2, A_3, A_4 and A_5; these points determine $\binom{5}{2} = 10$ lines connecting them. Fixing A_i, the number of lines joining the remaining four points is equal to $\binom{4}{2} = 6$, hence there are 6 orthogonals starting from A_i, from the 5 given points this adds up to 30 lines.

The maximal number of points of intersections of these 30 lines is $\binom{30}{2} = 435$.

Let us now determine how much this quantity *has to be* reduced because of possible coincidences and by disregarding the 5 given points.

a) The three orthogonals to $A_i A_k$ are parallel, so we have to reduce the number of points of intersection for each pair of points by 3, meaning the deduction of 30.

b) There are $\binom{5}{3} = 10$ triangles with vertices from the A_i's; there are three lines concurring in the orthocentres of these triangles. For this reason each triangle reduces by 2, altogether by 20 the number of points of intersection.

c) There are 6 orthogonals passing through a given point A_i, the $\binom{6}{2} = 15$ points of intersection of these lines do not count, hence we have to reduce the total number of points of intersection by $5 \cdot 15 = 75$.

The sum of the above reductions is $30 + 20 + 75 = 125$, consequently there are at most $435 \perp 125 = 310$ points of intersection.

Remark. The above reasoning did not show that the maximal number of points of intersection is equal to 310; for this we should provide a set of 5 points for which the number of appropriate intersection points of the orthogonals is eventually equal to 310. This can be done, but the description is very time- and space-consuming.

The original form of the problem — as it was given in the competition — went as follows: Determine the maximum number of intersections .... Right before the competition the members of the jury realized that the full solution would consume too much time, hence they verbally instructed the competitors that "the verification of maximum is not needed". At the end of the day, they expected roughly the same amount of work we presented in the solution above.

1964/6. *Let $ABCD$ be a given tetrahedron and D_1 the centroid of the face ABC. The parallels to DD_1 passing through the vertices A, B and C intersect the opposite faces in A_1, B_1 and C_1, respectively.*

a) *Show that the volume of $ABCD$ is one-third the volume of $A_1B_1C_1D_1$.*

b) *Is the result valid for any choice of D_1 in the interior of ABC?*

Solution. a) and b) will be shown at once; we will show that the volume of $A_1B_1C_1D_1$ is three times the volume of $ABCD$ for any choice of D_1.

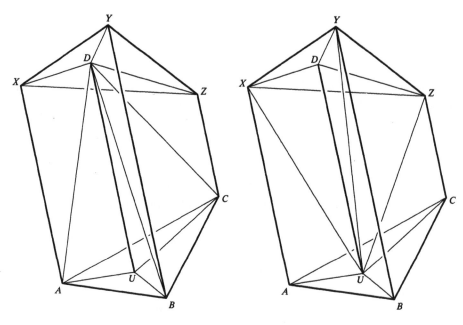

Figure 64/6.1a   *Figure 64/6.1b*

First we show that if the tetrahedrons $ABCD$ and $XYZU$ are such that the lines AX, BY, CZ and DU are parallel, D and U are interior points of the triangles XYZ and ABC, moreover ABC and XYZ are disjoint, then $ABCD$ and $XYZU$ have equal volumes (*Figures 1964/6.1a and 1b*).

Notice that the 4-sided pyramids $ABYXD$ and $ABYXU$ have the same base and the distances of D and U from $ABYX$ are equal. Therefore the volumes of $ABYXD$ and $ABYXU$ are equal. The same can be said for the pyramids $BCZYD$ and $BCZYU$, and for the pyramids $CAXZD$ and $CAXZU$. Now if we take away the first of the pyramids of the above pairs from the ungula of a prism $ABCXYZ$, we get the tetrahedron $ABCD$. By taking away the second of the pyramid pairs from $ABCXYZ$, we are left with the tetrahedron $XYZU$. This clearly implies that $ABCD$ and $XYZU$ have equal volumes.

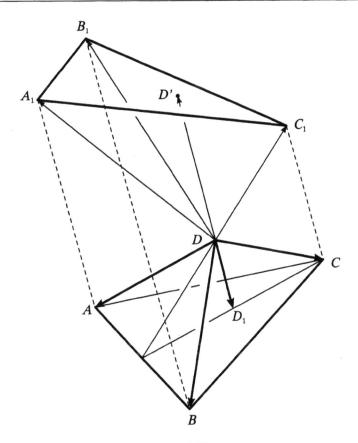

Figure 64/6.2

Now we turn to the solution of the problem. Fix D_1 in ABC and suppose that DD_1 intersects $A_1B_1C_1$ in D' (*Figure 1964/6.2*). According to our result above the tetrahedrons $ABCD'$ and $A_1B_1C_1D_1$ have equal volumes. If we prove that D_1D' is three times as much as DD_1 then we verified that the volume of $ABCD$ is third of the volume of $ABCD'$, hence of the volume of $A_1B_1C_1D_1$. In conclusion it is enough to show that $DD' = 2DD_1$.

We will use vector calculus in the verification of this last statement. Let D be the origin; vectors pointing to given points will be labelled by lower case letters coinciding with the labels of the points. Since D_1 is in the plane given by A, B and C, we get

(1) $\mathbf{d}_1 = \alpha\mathbf{a} + \beta\mathbf{b} + \gamma\mathbf{c},$

where $\alpha + \beta + \gamma = 1$, and this decomposition is unique (see [14]). Since CC_1 and DD_1 are parallel, we have $\mathbf{c}_1 = \mathbf{c} + \lambda\mathbf{d}_1$. Substitute (1) and conclude

(2) $\mathbf{c}_1 = \lambda\alpha\mathbf{a} + \lambda\beta\mathbf{b} + (1 + \lambda\gamma)\mathbf{c}.$

Since $\mathbf{c}_1$ is in the plane of $\mathbf{a}$ and $\mathbf{b}$, the coefficient of $\mathbf{c}$ is 0, hence $\lambda = -\dfrac{1}{\gamma}$.

Substituting this into (2) we get:

$$\gamma\mathbf{c}_1 = -\alpha\mathbf{a} - \beta\mathbf{b}.$$

Similar reasoning gives
$$\alpha\mathbf{a}_1 = \quad\; -\beta\mathbf{b} - \gamma\mathbf{c},$$
$$\beta\mathbf{b}_1 = -\alpha\mathbf{a} \quad\quad\; -\gamma\mathbf{c}.$$

The three equalities above yield

$$\alpha\mathbf{a}_1 + \beta\mathbf{b}_1 + \gamma\mathbf{c}_1 = -2\mathbf{d}_1.$$

Since $\alpha + \beta + \gamma = 1$, the endpoint of the vector on the left hand side is in the plane of $A_1 B_1 C_1$, while the endpoint of the one on the right hand side is on DD_1. This shows that

$$\mathbf{d}' = -2\mathbf{d}_1,$$

implying $DD' = 2DD_1$, which concludes the proof.

In summary, the volume of the tetrahedron $ABCD$ is third of the volume of $A_1 B_1 C_1 D_1$ for any allowable choices of D_1.

1965.

1965/1. *Find all x in the interval $[0, 2\pi]$ which satisfy*

(1) $$2\cos x \le |\sqrt{1 + \sin 2x} - \sqrt{1 - \sin 2x}| \le \sqrt{2}.$$

Solution. If x satisfies the inequality then it necessarily satisfies $\cos x \le$
$\le \dfrac{\sqrt{2}}{2}$, implying

(2) $$\frac{\pi}{4} \le x \le \frac{7\pi}{4}.$$

Let us first examine the right hand side inequality of (1):

$$|\sqrt{1 + \sin 2x} - \sqrt{1 - \sin 2x}| \le \sqrt{2}.$$

Since both sides are nonnegative, it is equivalent to its square

$$2 - 2\sqrt{1 - \sin^2 2x} \le 2,$$

i.e. to
$$-\sqrt{1 - \sin^2 2x} \le 0.$$

This latter expression is satisfied for all x since the left hand side is nonnegative.

Next we study the left hand side inequality of (1):

(3) $$2\cos x \le |\sqrt{1+\sin 2x} \mp \sqrt{1\mp\sin 2x}|.$$

This is obviously satisfied if $\cos x$ is nonpositive since the right hand side is nonnegative. Consequently

(4) $$\frac{\pi}{2}\le x \le \frac{3\pi}{2}$$

belongs to the solution set. For this reason it is enough to consider (3) on the intervals $\dfrac{\pi}{4}\le x\le\dfrac{\pi}{2}$ and $\dfrac{3\pi}{2}\le x\le\dfrac{7\pi}{4}$. Here both sides of (3) are nonnegative, hence (3) is equivalent to its square

$$4\cos^2 x \le 2 \mp 2\sqrt{1\mp\sin^2 2x}=2\mp 2|\cos 2x|,$$

from which $$2\cos^2 x \mp 1 \le \mp|\cos 2x|$$

and so $$\cos 2x \le \mp|\cos 2x|$$

follows. This is true if $\cos 2x \le 0$, giving the constraints $\dfrac{\pi}{2}\le 2x \le \dfrac{3\pi}{2}$ or $\dfrac{\pi}{2}+2\pi \le 2x \le \dfrac{3\pi}{2}+2\pi$ for x. These latter conditions transform to

$$\frac{\pi}{4}\le x\le\frac{3\pi}{4}\quad\text{or}\quad\frac{5\pi}{4}\le\frac{7\pi}{4}.$$

In summary, the solution set of (1) is the interval

$$\frac{\pi}{4}\le x\le\frac{7\pi}{4}.$$

1965/2. *The coefficients of the system of equations*

$$a_{11}x_1+a_{12}x_2+a_{13}x_3=0,$$
$$a_{21}x_1+a_{22}x_2+a_{23}x_3=0,$$
$$a_{31}x_1+a_{32}x_2+a_{33}x_3=0$$

are subject to the following constraints:
   a) a_{11}, a_{22} *and* a_{33} *are all positive,*
   b) *all other coefficients are negative,*
   c) *the sum of coefficients in each equation is positive.*

Verify that the only solution of the system is

$$x_1=x_2=x_3=0.$$

Solution. Suppose that the system admits a solution such that not all $x_i=0$. Since for a solution (x_1, x_2, x_3) also $(\mp x_1, \mp x_2, \mp x_3)$ is a solution, we may assume that one of the x_i's is positive, suppose that the largest one is x_1. Consider the first equation and notice that $x_2 \le x_1$ implies $a_{12}x_2 \ge a_{12}x_1$, so:

$$0 = a_{11}x_1 + a_{12}x_2 + a_{13}x_3 \ge a_{11}x_1 + a_{12}x_1 + a_{13}x_1 =$$
$$(a_{11} + a_{12} + a_{13})x_1 > 0.$$

This is, however, a contradiction, so no solution x_i can be different from 0.

Remarks. 1. Both the problem and the solution admits a (suitable) generalization for a system with n unknowns.

2. The solution also follows from the theorem asserting that such a system admits a nonzero solution if and only if its determinant vanishes. Hence by showing that this determinant is positive, we conclude $x_1 = x_2 = x_3 = 0$.

For this reason add the second and third columns to the first one of the determinant (these operations keep the value of the determinant unchanged):

$$D = \begin{vmatrix} q_1 & a_{12} & a_{13} \\ q_2 & a_{22} & a_{23} \\ q_3 & a_{32} & a_{33} \end{vmatrix},$$

here q_1, q_2 and q_3 are the sums of the coefficients in the rows, respectively, hence these are positive. Now we get $D = q_1(a_{22}a_{33} \perp a_{23}a_{32}) + q_2(a_{32}a_{13} \perp a_{12}a_{33}) + q_3(a_{12}a_{23} \perp a_{13}a_{22})$, and in this expression the second and the third summands are positive because of a) and b) respectively. $a_{22} + a_{23} > a_{21} + a_{22} + a_{23} > 0$ implies $a_{22} > \perp a_{23}$, and similarly $a_{33} > \perp a_{32}$. Therefore $a_{22}a_{33} > a_{23}a_{32}$, and so the first summand, and so D is positive.

1965/3. *The length of the edge AB in the tetrahedron $ABCD$ is a, while the length of CD is b. The distance between the skew lines AB and CD is d, the angle determined by them is ω. The tetrahedron is divided into two parts by a plane ε parallel to AB and CD. We also know that k times the distance between AB and ε equals the distance between CD and ε. Determine the ratio of the volumes of the parts of the tetrahedron.*

Solution. In order to get a better feeling about the problem, we pictured the tetrahedron on *Figure 1965/3.1* by the parallelepiped containing it (see [15]). Now ε cuts out the parallelogram $XYZU$ from it if we intersect two planes by a third plane parallel to their intersection line then the resulting intersection lines will be parallel.

According to the problem the distance between the lower and the upper faces of the parallelepiped is d, therefore the distance of ε from the lower face is $\frac{d}{k+1}$, and from the upper face this distance is $\frac{kd}{k+1}$. Let $XY = ZU = a'$ and

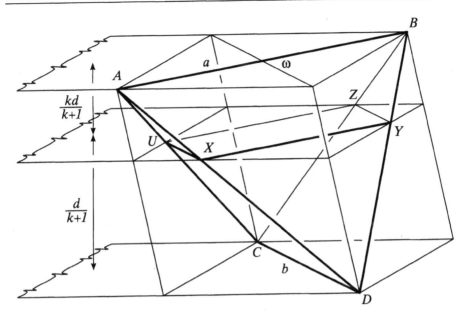

Figure 65/3.1

$YZ = UX = b'$. By shrinking the edge AB from D with quotient $XY : AB =$
$= \dfrac{d}{k+1} : d = \dfrac{1}{k+1}$ we get XY, hence

(1) $\qquad\qquad\qquad XY = a' = \dfrac{a}{k+1}, \quad \dfrac{a}{a'} = k+1.$

Similarly, shrinking CD from B with quotient $\dfrac{dk}{k+1} : d = \dfrac{k}{k+1}$ we get

(2) $\qquad\qquad\qquad YZ = b' = \dfrac{bk}{k+1}, \quad \dfrac{b}{b'} = \dfrac{k+1}{k}.$

The plane ε divides the tetrahedron into the solids $ABXYZU$ and $CDXYZU$; the area of their common base will be denoted by t while their altitudes are $\dfrac{kd}{k+1}$ and $\dfrac{d}{k+1}$. Their edges parallel to XY and YZ are a and b, finally their volumes are V_1 and V_2 respectively. Using (1) and (2) the ratio of the volumes is

$$V_1 : V_2 = \frac{kdt}{6(k+1)} \left(2 + \frac{a}{a'} \right) : \frac{dt}{6(k+1)} \left(2 + \frac{b}{b'} \right) = \frac{k(k+3)}{\frac{3k+1}{k}} = \frac{k^2(k+3)}{3k+1}.$$

Remarks. 1. The solution used a formula expressing the volume of the solids $ABXYZU$ and $CDXYZU$. Such solids (also called roof-shapes) can be got by intersecting an (infinite) triangular prism with two planes in such a way that the remainder of one face is a parallelogram, becoming the base of the solid (on *Figure 1965/3.2* this is the parallelogram $XYZU$). The edge parallel to the base

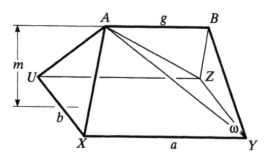

Figure 65/3.2

face is called the spine (denoted by $AB = g$ here). The distance of the spine from the base is the height m of the solid. The sides of the base will be denoted by a and b ($a \| g$). The solid $ABXYZU$ can be decomposed into the pyramid $AXYZU$ and the tetrahedron $ABYZ$. Notice that the angle ω of the sides a and b of the base parallelogram is equal to the angle of the edges AB and YZ. Since the volume of the roof-shaped solid is simply the sum of the volumes of the pyramid and the tetrahedron, we get

$$V = \frac{mab \sin \omega}{3} + \frac{mgb \sin \omega}{6} = \frac{mab \sin \omega}{6}\left(2 + \frac{g}{a}\right).$$

If the area of the base parallelogram is t (so $t = ab \sin \omega$), the above equality implies

$$V = \frac{tm}{6}\left(2 + \frac{g}{a}\right).$$

2. Notice that the ratio of the volumes depends only on k. The reason for this phenomenon is that two tetrahedrons can be transformed into each other by an affine transformation. Such a transformation maps parallel objects to parallel ones and preserves the ratio of collinear intervals and volumes. Therefore the ratio should be the same for all tetrahedrons.

1965/4. *Find all sets of four real numbers x_1, x_2, x_3, x_4 such that the sum of any one and the product of the other three is 2.*

Solution. The problem is equivalent to solving the system

$$x_1 + x_2 x_3 x_4 = 2,$$
$$x_2 + x_3 x_4 x_1 = 2,$$
$$x_3 + x_4 x_1 x_2 = 2,$$
$$x_4 + x_1 x_2 x_3 = 2.$$

Notice first that $x_i \neq 0$, since for $x_1 = 0$ the last three equations give $x_2 = x_3 = = x_4 = 2$ which do not satisfy the first equation. For a solution (x_1, x_2, x_3, x_4) let

Q denote the product of the x_i's. It is easy to see that all four x_i satisfies

$$x + \frac{Q}{x} = 2.$$

This means that all x_i are roots of $x^2 - 2x + Q = 0$, hence the possible values of x_i are

$$1 + \sqrt{1 - Q} \quad \text{and} \quad 1 - \sqrt{1 - Q}.$$

We conclude that there are at most two different values among the x_i's. In the following we will analyze the various cases.

a) All x_i are equal. Let $x_i = k$ ($i = 1, 2, 3, 4$), then $k + k^3 = 2$, so

$$(k - 1)(k^2 + k + 2) = 0.$$

Since the second term has no real roots, the only such solution is $x_1 = x_2 = x_3 = x_4 = k = 1$ which solves the system.

b) Three x_i's are equal and the fourth is different from them. Let $x_1 = x_2 = x_3 = k$, $x_4 = n$ and $k \neq n$. The first and the fourth equations now read as

$$k + k^2 n = 2$$
$$n + k^3 = 2.$$

The difference of the two equations imply $(k - n)(1 - k^2) = 0$; since $k \neq n$, this gives $k^2 = 1$ and so $k = \pm 1$. For $k = 1$ the fourth equation gives $n = k$ which is excluded. In case $k = -1$ the fourth equation implies

$$n - 1 = 2, \qquad n = 3,$$

providing $x_1 = x_2 = x_3 = -1$, $x_4 = 3$, and all permutations of these. These numbers satisfy the original system of equations.

c) Finally we assume that $x_1 = x_2 = k$, $x_3 = x_4 = n$ and $k \neq n$. The first and the fourth equations give

$$k + kn^2 = 2,$$
$$n + k^2 n = 2,$$

and their difference shows $(k - n)(1 - nk) = 0$. Since $k \neq n$, we have $nk = 1$. Substituting this we get $k + n = 2$, hence n and k are two different solutions of $x^2 - 2x + 1 = (x - 1)^2 = 0$. Since this equation provides a unique solution $1 = k = n$, we conclude that c) does not provide further solutions.

In summary, the solutions are

$$(1, 1, 1, 1); \ (3, -1, -1, -1); \ (-1, 3, -1, -1); \ (-1, -1, 3, -1); \ (-1, -1, -1, 3).$$

1965/5. *The triangle OAB has angle $\angle AOB$ acute. M is an arbitrary point in OAB different from O. The points P and Q are the feet of the perpendiculars from M to OA and OB, respectively. Determine the locus of the orthocentre H of the triangle OPQ if M is*

a) *on AB;*

b) *in the interior of OAB.*

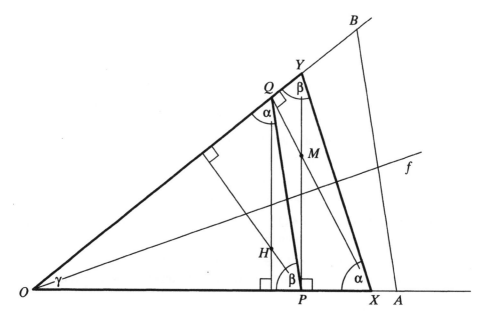

Figure 65/5.1

Solution. Suppose that M is not on the lines OA and OB. Let Y and X denote the points of intersection of MP with OB and of MQ with OA (*Figure 1965/5.1*). Since M is the orthocentre of the triangle OXY, the correspondence $M \leftrightarrow H$ associates the orthocentre of OXY to the orthocentre of OQP. Let us first examine this correspondence.

Since X, Y, Q, P are on the same circle (hence both XPY and XQY are right angles), $XYPQ$ admits a circumcircle, hence

$$\angle OXY = \angle OQP = \alpha, \qquad \angle OYX = \angle OPQ = \beta.$$

Therefore the triangles OXY and OQP are similar with ratio

$$\frac{OP}{OY} = \cos \gamma,$$

where γ is the given acute angle. Consequently, by shrinking OXY from O with $\cos \gamma$ and then reflecting it to the bisector f of $\angle AOB$ we get OQP. It means that the correspondence $M \leftrightarrow H$ is a similarity with centre O, axis f and ratio $\cos \gamma$. It is easy to see that the above said holds in case M happens to be on OA or on OB. Consequently, the image of the triangle OAB (and so the interval AB contained by it) can be got by shrinking it with ratio $\cos \gamma$ and then reflecting it to f. Since this transformation is one-to-one, the locus is simply the image of the interval AB (the triangle OAB) under the above transformation.

1965/6. *For $n \geq 3$ points in the plane denote the maximal distance of pairs of points by d. Prove that at most n pairs of points are of distance d apart.*

Solution. Let the points be denoted by $P_1, P_2, \ldots, P_{n-1}$. An interval of length d joining two points P_i and P_j is called the diameter of the set. The degree of a point is by definition the number of diameters starting from it. Since the sum of the degrees is twice the number of diameters, in case the degree of each point is at most 2 we get that the sum of degrees is at most $2n$, yielding that there are at most n diameters.

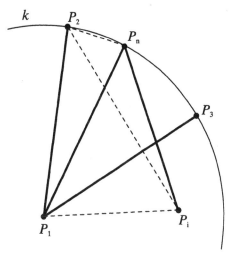

We will solve the problem by induction on n: it is obviously true if $n = 3$ and if all points $P_1, P_1, \ldots, P_n$ have degrees at most two. Suppose now that there is a point, say P_1, of degree at least 3 (*Figure 1965/6.1*). According to the inductive hypothesis, among the remaining $n \perp 1$ points there are at most $n \perp 1$ diameters.

The diameters starting from P_1 are on an arc of a circle k of radius d around P_1. Let the extremal points on this arc be called P_2 and P_3. As $P_2 P_3$ is not more than d, the arc $P_2 P_3$ is not more than $60°$. Since the degree of P_1 is at least three, there is a diame-

Figure 65/6.1

ter joining it with an inner point of the arc $P_1 P_2$; we denote this point by P_n. Next we show that the degree of P_n is one.

Suppose to the contrary that there is another diameter $P_n P_i$ (besides $P_n P_1$) starting from P_n. Obviously P_i is not on k, and cannot be outside of k, since then $P_1 P_i$ is larger than d. In case P_i is inside k, it cannot lie in the triangle $P_1 P_2 P_3$ since this implies $P_n P_i < d$; consequently P_i is outside of $P_1 P_2 P_3$. This implies that $P_n P_1$ intersects $P_1 P_2$ or $P_1 P_3$; without loss of generality we can assume that it intersects $P_1 P_3$. This shows that $P_1 P_2 P_n P_i$ is a convex quadrilateral, and according to a famous inequality (see Remark 2.) the sum of the diagonals is more than the sum of any pair of opposite sides:

$$P_1 P_2 + P_n P_i < P_1 P_n + P_2 P_i, \quad \text{i.e.,} \quad d < P_2 P_i.$$

This, however, contradicts the definition of d, hence P_i does not exist, so the degree of P_n is one. Let us now delete P_n; according to the inductive hypothesis there are only $n \perp 1$ diameters on the remaining $n \perp 1$ points, and now adding P_1 back we get the statement we wanted to prove.

Remarks. 1. The basic combinatorial geometric theorem on which the problem rests is due to *H. Hopf* and *E. Panwitz*, and there are many related problems to it.

2. We used the following result: if $ABCD$ is a convex quadrilateral then $AB+CD < AC+BD$. Let M denote the point of intersection of AC and BD. By summing the triangle inequalities $AB < AM+BM$ and $CD < CM+DM$ we get the desired statement.

3. The problem admits a generalization into the three-space: $n > 3$ points in the space determine at most $2n \perp 2$ diameters.

1966.

1966/1. *Problems A, B and C have been posed in a mathematical contest. 25 competitors solved at least one of the three. Amongst those who did not solve A, twice as many solved B as C. The number of competitors solving only A was one more than the number of competitors solving A and at least one other problem. The number of competitors solving A equalled the number solving just B plus the number of competitors solving just C. How many competitors solved just B?*

Solution. Let us visualize the sets of competitors solving A, B or C by discs, while the students solving more problems correspond to intersections of these discs (*Figure 1966/1.1*).

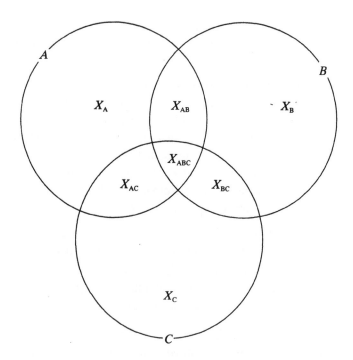

Figure 66/1.1

Now let X_A, X_B, X_C denote the number of those competitors who solved *just A, B or C*, X_{AB} denotes the number of those who solved *just A and B*, and likewise X_{ABC} stands for the number of students solving all three problems. According to the problem

$$X_A + X_B + X_C + X_{AB} + X_{BC} + X_{AC} + X_{ABC} = 25,$$
$$X_B + X_{BC} = 2(X_C + X_{BC}),$$
$$X_A = 1 + X_{AB} + X_{AC} + X_{ABC},$$
$$X_A + X_B + X_C = 2(X_B + X_C).$$

We have to determine the (positive integer) X_B. After ordering we have:

(1) $\qquad X_A + X_B + X_C + X_{AB} + X_{BC} + X_{AC} + X_{ABC} = 25,$
(2) $\qquad X_B \perp 2X_C \qquad \perp X_{BC} \qquad\qquad\qquad = 0,$
(3) $\qquad X_A \qquad\qquad \perp X_{AB} \qquad \perp X_{AC} \perp X_{ABC} = 1,$
(4) $\qquad X_A \perp X_B \perp X_C \qquad\qquad\qquad\qquad = 0.$

Now subtract twice the fourth equation from the sum of the first three and get:

(5) $$4X_B + X_C = 26.$$

From (2) it follows that $X_B \perp 2X_C = X_{BC} \geq 0$, hence $X_B \geq 2X_C$, therefore (5) implies

$$26 = 4X_B + X_C \geq 8X_C + X_C = 9X_C.$$

Consequently

$$X_C \leq \frac{26}{9},$$

and since X_C is a positive integer, its possible values are 0, 1 and 2. For $X_C = 0$ and 1 equation (5) does not admit an integer solution for X_B, hence the only possibility remains $X_C = 2$. This implies

$$X_B = 6,$$

hence B has been solved by six competitors.

Remark. From the above solution now it is easy to determine the remaining unknowns $X_A = 8$ and $X_{BC} = 2$; the remaining three unknowns, however, can take any positive integer values with sum equal to 7:

$$X_{AB} + X_{AC} + X_{ABC} = 7.$$

1966/2. *Let a, b, c denote the sides of a triangle, while the opposite angles are denoted by* α, β, γ. *Prove that if*

(1) $$a + b = \tan \frac{\gamma}{2}(a \tan \alpha + b \tan \beta)$$

then the triangle is isosceles.

First solution. We assume that the expression under (1) does make sense; now express the tangent function as the quotient of sine and cosine and multiply the equation by the product of the denominators. After ordering we get

$$a+b=\frac{\sin\frac{\gamma}{2}}{\cos\frac{\gamma}{2}}\left(\frac{a\sin\alpha}{\cos\alpha}+\frac{b\sin\beta}{\cos\beta}\right).$$

$$a\cos\beta\left(\cos\alpha\cos\frac{\gamma}{2}\perp\sin\alpha\sin\frac{\gamma}{2}\right)+b\cos\alpha\left(\cos\beta\cos\frac{\gamma}{2}\perp\sin\beta\sin\frac{\gamma}{2}\right)=0,$$

$$a\cos\beta\cos\left(\alpha+\frac{\gamma}{2}\right)+b\cos\alpha\cos\left(\beta+\frac{\gamma}{2}\right)=0.$$

Notice that since

$$\left(\alpha+\frac{\gamma}{2}\right)+\left(\beta+\frac{\gamma}{2}\right)=180°,$$

we have

$$\cos\left(\beta+\frac{\gamma}{2}\right)=\perp\cos\left(\alpha+\frac{\gamma}{2}\right),$$

hence (1) is equivalent to

(2) $$\cos\left(\alpha+\frac{\gamma}{2}\right)(a\cos\beta\perp b\cos\alpha)=0.$$

If $\cos\left(\alpha+\frac{\gamma}{2}\right)=0$ then $\alpha+\frac{\gamma}{2}=90°$, and also $\beta+\frac{\gamma}{2}=90°$ is satisfied. This shows that $\alpha=\beta$, consequently the triangle is isosceles.

If the second term vanishes then

$$a\cos\beta=b\cos\alpha,$$

i.e.,

(3) $$a^2\cos^2\beta=b^2\cos^2\alpha.$$

Add the square of the equation $a\sin\beta=b\sin\alpha$ (originated from the law of sines) to (3) and get

$$a^2\sin^2\beta=b^2\sin^2\alpha,$$
$$a^2=b^2,$$

showing again that the triangle is isosceles.

Second solution. There are various geometric interpretations of (1). Let us consider those two right angled triangles for which one acute angle is α (and β, respectively), and one adjacent side is a (and b, respectively). The opposite sides in these triangles are equal to $a\tan\alpha$ and $b\tan\beta$, respectively. Put these two triangles next to each other as shown by *Figure 1966/2.1.* Now inscribe the two triangles into a right angled triangle ABC with sides $a+b$ and $a\tan\alpha+b\tan\beta$. (Since a and b have symmetric roles, we may assume that $b\geq a$.)

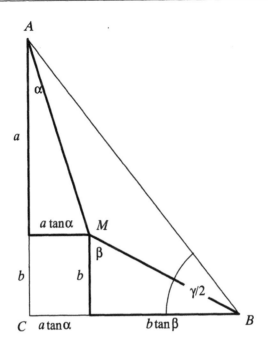

Figure 66/2.1

Since (1) uniquely determines $\dfrac{\gamma}{2}$, and in the triangle ABC we have

$$\tan \angle ABC = \frac{a+b}{a \tan \alpha + b \tan \beta},$$

we get that $\angle ABC = \dfrac{\gamma}{2}$. If the common vertex of our original subtriangles is denoted by M, we have

$$\angle MAB = 90° \perp \frac{\gamma}{2} \perp \alpha, \quad \text{and so} \quad \angle MAB = \alpha \perp \left(90° \perp \frac{\gamma}{2}\right) = \alpha \perp 90° + \frac{\gamma}{2},$$

$$\angle MBA = \frac{\gamma}{2} \perp (90° \perp \beta), \quad \text{and so} \quad \angle MBA = 90° \perp \beta \perp \frac{\gamma}{2},$$

depending on whether M is on the same side of AB as C or it is on the opposite side. In both cases:

$$|\angle MAB \perp \angle MBA| = |180° \perp \alpha \perp \beta \perp \gamma| = 0, \quad \text{hence} \quad MAB\angle = \angle MBA.$$

Therefore the (possibly degenerate) triangle ABM is isosceles, hence $AM^2 = {} = BM^2$. This shows that in the subtriangles

$$a^2(1 + \tan^2 \alpha) = b^2(1 + \tan^2 \beta),$$
$$a^2 \cos^2 \beta = b^2 \cos^2 \alpha,$$

which is identical to equation (3) in our first solution. Since (3) implied $a = b$, the statement follows.

Remark. In the second solution it has been used that α and β are acute. This is legitimate, since the notation can be chosen to achieve $b \geq a$, hence $\beta \geq$ $\geq \alpha$. This shows that β cannot be obtuse, since in this case $180° \perp \beta$ is acute, implying

$$90° > 180° \perp \beta = \alpha + \gamma > \alpha,$$

and therefore $\perp \tan \beta > \tan \alpha$, i.e., (since $\tan \beta$ is negative) $\perp \tan \beta = |\tan \beta| >$ $> \tan \alpha$ and (from $b \geq a$) also $|b \tan \beta| > a \tan \alpha$. This, however, shows that the right hand side of (1) is negative, which is impossible since $a + b$ is positive. In conclusion β and so α are both acute.

1966/3. *Prove that a point in the space has the smallest sum of distances to vertices of a regular tetrahedron if and only if it is the centre of the tetrahedron.*

First solution. For a better picture we draw the regular tetrahedron $ABCD$ with the cube containing it ([15]), see *Figure 1966/3.1.* The line t joining the midpoints of the edges AB and CD is orthogonal to the two opposite faces of

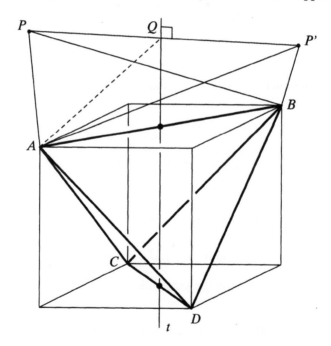

Figure 66/3.1

the cube, hence t is orthogonal to the edges AB and CD.

Let P be an arbitrary point in the space. The reflection of P on t is P' and the midpoint of PP' — i.e., the foot of the orthogonal from P to t — is Q. Let us first study the sum of the distances of P from A and B, which is $PA + PB$. Since we applied reflection, we get $PB = P'A$ and so $PA + PB = PA + P'A$.

The interval QA is a median of PAP'. According to the median inequality (see Remark 2.), the median of a triangle is shorter than the arithmetic mean of the two adjacent sides, therefore the sum of the distances from Q to A and B is less than the same sum from P once P is not on t.

The same holds for the distances from C and D (since we only used the fact that the reflection of A to t is exactly B, and the same relation holds for C and D). Therefore if P is not on t then it cannot minimize the sum of distances from the vertices of $ABCD$.

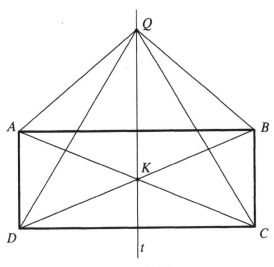

Figure 66/3.2

The distance of Q from A and B remains unchanged if we rotate AB around t until it becomes parallel to CD. In this case t becomes the median of the rectangle $ABCD$. Let K denote the centre of this rectangle. The sum of the distances between K and the vertices A, B, C, D is less than the same sum for any other point of t, since if Q is a point of t different from K (*Figure 1966/3.2*), then according to the triangle inequality $QA + QC > AC$ and $QB + QD > BD$, showing

$$QA + QB + QC + QD > KA + KB + KC + KD.$$

Now K is the centre of the circumsphere of the tetrahedron $ABCD$, hence we showed that the sum of distances from K is smaller than the corresponding sum for any other point in the space.

Second solution. This solution rests on the following statement: the sum of (oriented) distances of any point P in the space from the faces of a regular tetrahedron is constant, moreover it is equal to the altitude of the tetrahedron. (The distance of a point from a face of the tetrahedron is positive if it is on the same side of the face as the opposite vertex. For example, for inner points of the tetrahedron all distances are positive.)

The proof of this statement proceeds by computing volumes. Let the distance of P from the face ABC of the regular tetrahedron $ABCD$ be negative, and positive from all other faces (*Figure 1966/3.3*). The tetrahedrons $PABD$, $PBCD$ and $PACD$ together provide a cover of $ABCD$, and even more; this

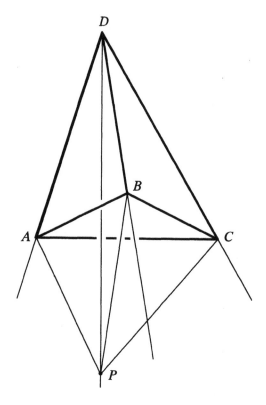

Figure 66/3.3

additional piece is exactly the tetrahedron $PABC$. Let us denote the area of the faces by t, the altitudes by m and the distances of P from the various faces by d_A, d_B, d_C and d_D (in our case d_D is negative). Since

$$V_{ABCD} = V_{PABD} + V_{PBCD} + V_{PACD} \perp V_{PABC},$$

multiplying it by 3 we get

$$tm = td_C + td_A + td_B + td_D,$$

implying

$$d_A + d_B + d_C + d_D = m \quad \text{(constant)}.$$

A similar method works for points in other space segments, the number of negative distances varies from 1 to 3.

Now we are ready to prove the statement of the problem. For a point P in the space d_A, d_B, d_C, d_D denote its distances from the faces, and the feet of the orthogonals are denoted by P_A, P_B, P_C and P_D, respectively. The triangle inequality implies

$$PA + PP_A \geq m, \quad \text{i.e.} \quad PA + d_A \geq m,$$

since $PA + PP_A \geq AP_A \geq m$. This last inequality remains true even if $PP_A = d_A$ is negative (*Figure 1966/3.4*), with equality if and only if P is on the altitude

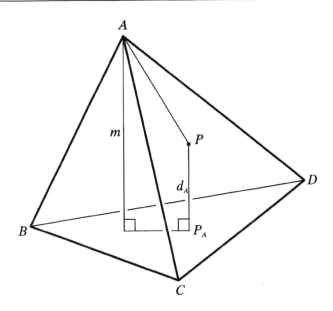

Figure 66/3.4

starting from A. Writing down the above inequalities for all vertices we get

$$PA+d_A \geq m, \quad PB+d_B \geq m, \quad PC+d_C \geq m, \quad PD+d_D \geq m.$$

By summing them we conclude

$$PA+PB+PC+PD \geq 4m \perp (d_A+d_B+d_C+d_D) = 3m$$

with equality if and only if P is on all altitudes of the tetrahedron, in other words, if P is the centre of the tetrahedron.

Remarks. 1. The first solution did not really made use of the regularity of the tetrahedron. We only used that it admits two edges with equal length and the interval joining their midpoints is perpendicular to both. This property is satisfied by a wider class of tetrahedrons.

2. The proof of the median inequality used in the first solution can be verified as follows: Let the reflection of the vertex A of the triangle ABC to the midpoint of the opposite edge be denoted by A'. Now $AA' = 2s_a$, and the triangle inequality for ABA' gives $b+c > 2s_a$, or equivalently $s_a < \dfrac{b+c}{2}$.

1966/4. *Prove that*

(1)
$$\frac{1}{\sin 2x} + \frac{1}{\sin 4x} + \ldots + \frac{1}{\sin 2^n x} = \cot x \perp \cot 2^n x.$$

for any positive integer n and any real x (with $x \neq \dfrac{\lambda \pi}{2^k}$, $k=0, 1, 2, \ldots, n$ and λ an arbitrary integer).

Solution. For such x we have the identity

$$\cot x \perp \cot 2x = \frac{\cos x}{\sin x} \perp \frac{\cos 2x}{\sin 2x} = \frac{\cos x}{\sin x} \perp \frac{\cos^2 x \perp \sin^2 x}{2 \sin x \cos x} = \frac{1}{\sin 2x}.$$

In verifying the equality of the problem we just use the above identity over and over again:

$$\frac{1}{\sin 2x} + \frac{1}{\sin 4x} + \ldots + \frac{1}{\sin 2^n x} = (\cot x \perp \cot 2x) + (\cot 2x \perp \cot 4x) + \ldots +$$
$$+ (\cot 2^{n-1} x \perp \cot 2^n x) = \cot x \perp \cot 2^n x,$$

which is exactly the statement we wanted to prove.

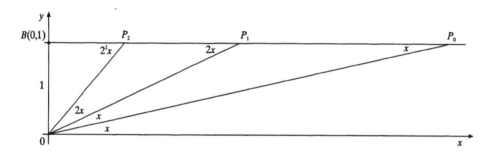

Figure 66/4.1

Remark. In case all the angles appearing are acute, the problem has a geometric interpretation. The line with equation $y = 1$ is intersected with the lines passing through the origin with slopes $x, 2x, \ldots, 2^n x$ in the points $P_0, P_1, \ldots$ $\ldots, P_n$ (*Figure 1966/4.1*); the distances of these points from $B(0,1)$ are $\cot x$, $\cot 2x$, ..., $\cot 2^n x$. Consequently

$$P_{i+1} P_i = \cot 2^i x \perp \cot 2^{i+1} x,$$

therefore twice the area of the triangle $OP_{i+1} P_i$ is $\cot 2^i x \perp \cot 2^{i+1} x$, while twice the area of $OP_{n+1} P_1$ is equal to $\cot x \perp \cot 2^n x$.

On the other hand, $OP_i = \dfrac{1}{\sin 2^i x}$ and $OP_{i+1} = \dfrac{1}{\sin 2^{i+1} x}$, hence twice the area of $OP_{i+1} P_i$ is

$$\frac{\sin 2^i x}{\sin 2^i x \cdot \sin 2^{i+1} x} = \frac{1}{\sin 2^{i+1} x}.$$

Now (1) just says that the area of $OP_{n+1} P_1$ is the sum of the subtriangles $OP_{i+1} P_i$.

1966/5. *Solve the system of equations*

$$|a_1 \perp a_2|x_2 + |a_1 \perp a_3|x_3 + |a_1 \perp a_4|x_4 = 1,$$
$$|a_2 \perp a_1|x_1 \qquad\qquad + |a_2 \perp a_3|x_3 + |a_2 \perp a_4|x_4 = 1,$$
$$|a_3 \perp a_1|x_1 + |a_3 \perp a_2|x_2 \qquad\qquad + |a_3 \perp a_4|x_4 = 1,$$
$$|a_4 \perp a_1|x_1 + |a_4 \perp a_2|x_2 + |a_4 \perp a_3|x_3 \qquad\qquad = 1,$$

where a_1, a_2, a_3, a_4 denote four distinct reals.

Solution. First we will get rid of the absolute value signs. Notice that by permuting the a_i's the system remains unchanged, so assume that $a_e < a_k < < a_h < a_n$ (here $e\ k\ h\ n$ is just a permutation of 1 2 3 4). Introduce the notations $c_1 = a_e$, $c_2 = a_k$, $c_3 = a_h$, $c_4 = a_n$ and $x_e = y_1$, $x_k = y_2$, $x_h = y_3$, $x_n = y_4$, hence $c_1 < < c_2 < c_3 < c_4$. Now the system has the following form:

(1) $$(c_2 \perp c_1)y_2 + (c_3 \perp c_1)y_3 + (c_4 \perp c_1)y_4 = 1,$$
(2) $$(c_2 \perp c_1)y_1 + \qquad\qquad (c_3 \perp c_2)y_3 + (c_4 \perp c_2)y_4 = 1,$$
(3) $$(c_3 \perp c_1)y_1 + (c_3 \perp c_2)y_2 + \qquad\qquad (c_4 \perp c_3)y_4 = 1,$$
(4) $$(c_4 \perp c_1)y_1 + (c_4 \perp c_2)y_2 + (c_4 \perp c_3)y_3 \qquad\qquad = 1.$$

By taking the difference of each equation with its preceding equation together with the sum of the first and the last we get

$$(c_2 \perp c_1)(y_1 \perp y_2 \perp y_3 \perp y_4) = 0,$$
$$(c_3 \perp c_2)(y_1 + y_2 \perp y_3 \perp y_4) = 0,$$
$$(c_4 \perp c_3)(y_1 + y_2 + y_3 \perp y_4) = 0,$$
$$(c_4 \perp c_1)(y_1 + y_2 + y_3 + y_4) = 2,$$

The c_i's are distinct, so we can divide by their differences, resulting

(5) $$y_1 \perp y_2 \perp y_3 \perp y_4 = 0,$$
(6) $$y_1 + y_2 \perp y_3 \perp y_4 = 0,$$
(7) $$y_1 + y_2 + y_3 \perp y_4 = 0,$$
(8) $$y_1 + y_2 + y_3 + y_4 = \frac{2}{c_4 \perp c_1}.$$

The difference of (6) and (5), and similarly the difference of (7) and (6) shows

$$y_2 = y_3 = 0;$$

and the difference of (8) and (7) together with the sum of (5) and (8) gives

$$y_4 = y_1 = \frac{1}{c_4 \perp c_1}.$$

This implies that in case $a_e < a_k < a_h < a_n$ the solution is

$$x_k = x_h = 0, \qquad x_e = x_n = \frac{1}{a_n \perp a_e}.$$

Now it is not hard to verify that this is actually a solution of our original system.

1966/6. *Take any points K, L, M on the sides AB, BC, CA of the triangle ABC. Prove that at least one of the triangles MAL, KBM and LCK has area at most fourth the area of ABC.*

First solution. In the following we will adopt the convention that if an interval of length d is cut into two pieces then one will be denoted by dx while the other one by $d(1 \perp x)$ $(0 < x < 1)$.

Let the sides of the triangle be denoted by c, a and b, and the points M, K, L divide them as cz, $c(1 \perp z)$; ax, $a(1 \perp x)$ and by, $b(1 \perp y)$, respectively (x, y, z are positive reals < 1, see *Figure 1966/6.1*). Let t denote the area of ABC, the area of the subtriangle MAL at A is t_A, while for the other two t_B and t_C denote their respective areas.

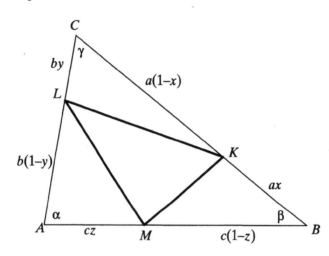

Figure 66/6.1

We need to show that among the ratios
$$\frac{t_A}{t}, \ \frac{t_B}{t}, \ \frac{t_C}{t}$$
there is one not exceeding $\frac{1}{4}$. Notice that
$$\frac{t_A}{t} = \frac{b(1 \perp y)cz \cdot \sin \alpha}{bc \sin \alpha} = (1 \perp y)z,$$
and similarly
$$\frac{t_B}{t} = (1 \perp z)x, \quad \frac{t_C}{t} = (1 \perp x)y.$$
Therefore we only need to demonstrate that the products

(1) $$(1 \perp y)z, \ \ (1 \perp z)x, \ \ (1 \perp x)y$$

contain one not more than $\frac{1}{4}$.

Apply the inequality between the arithmetic and geometric means and conclude that

$$x(1 \perp x) \le \left(\frac{x + 1 \perp x}{2}\right)^2 = \frac{1}{4}.$$

$$(1 \perp y)z \cdot (1 \perp z)x \cdot (1 \perp x)y = x(1 \perp x) \cdot y(1 \perp y) \cdot z(1 \perp z) \le \left(\frac{1}{4}\right)^3,$$

hence not all the terms in the product can be larger than $\frac{1}{4}$. This, however, concludes our solution.

Second solution. Let the midpoints of AB, BC, CA be denoted by C', A', B'. The sides of the subtriangle $A'B'C'$ partition ABC into four subtriangles with areas equal to fourth of the area of ABC.

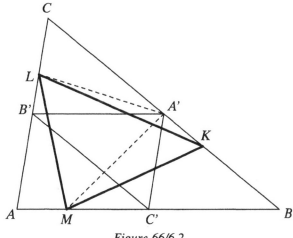

If two of the points K, L, M are in one of the subtriangles then the statement of the problem is obvious. Therefore we can assume (after possible renaming) that K, L and M are on the intervals BA', CB' and AC', respectively (*Figure 1966/6.2*). We will show that the area of KLM is at least fourth the area of ABC, in other words its area is at least the area of $A'B'C'$.

Figure 66/6.2

By pushing the vertex K of KLM into A' we do not increase the area of KLM since the altitude belonging to LM did not increase. Therefore

(2)   area of $KLM \ge$ area of $A'LM$.

Now the area of $A'LM$ does not increase when we push L into B', since the altitude belonging to MA' did not increase, hence

(3)   area of $A'LM \ge$ area of $A'B'M$.

Notice, however, that the area of $A'B'M$ coincides with the area of $A'B'C'$ (consequently with fourth the area of ABC), so (2) and (3) imply

$$\text{area of } KLM \ge \frac{1}{4} \cdot \text{area of } ABC.$$

Therefore the sum of the three subtriangles of ABC given by the sides of KLM is at most $\frac{3}{4}$ the area of ABC, hence there is one with area at most fourth the area of ABC.

Remarks. 1. Since (1) is unsensitive for cyclic permutations of x, y and z, we may assume that x is the largest among x, y, z, hence $y \le x$, showing

$$(1 \perp x)y \le (1 \perp x)x \le \frac{1}{4}.$$

2. There are many noteworthy connections between ABC, KLM and the further subtriangles; here we mention only a few of these

a) one of the triangles MAL, KBM and LCK has area not more than the area of KLM;

b) the circumference of one of MAL, KBM and LCK is not more than the circumference of KLM.

1967.

1967/1. *The parallelogram $ABCD$ has $AB = a$, $AD = 1$, angle $\angle DAB = \alpha$ and the triangle ABD is acute. Prove that the circles K_A, K_B, K_C and K_D of radius 1 centered at A B, C and D cover the parallelogram if and only if*

(1) $a \le \cos \alpha + \sqrt{3} \sin \alpha.$

Solution. Let R denote the radius of the circumcircle of ABD (and so of CDB, which is congruent to it). If K_A, K_B, K_D cover ABD, then K_B, K_C, K_D cover CDB. We will show that $R \le 1$ is a necessary condition for this to happen.

Since ABD is acute, the centre O of its circumcircle is in its interior. We obviously need $R \le 1$ in order to cover O. This assumption is also sufficient: in order to show this, choose P in ABD (*Figure 1967/1.1*).

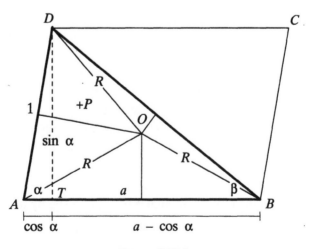

Figure 67/1.1

P is in one of the six right angled triangles we get by considering the radii OA, OB, OC together with the orthogonals from O to the sides. Now if P is in one of the triangles containing D as a vertex, then since $DP \le DO = R$, K_D covers P.

We only need to show that $R \le 1$ is equivalent to (1).

Let $\angle ABD = \beta$. Since $R = \dfrac{AD}{2 \sin \beta} = \dfrac{1}{2 \sin \beta}$, we get that $R \le 1$ is equivalent

to $\sin \beta \ge \dfrac{1}{2}$, i.e. to $\beta \ge 30°$. Let the foot of the orthogonal from D to AB be denoted by T. In DTB we have $DT = \sin \alpha$, $TB = a \perp \cos \alpha$ and so

$$\cot \beta = \frac{a \perp \cos \alpha}{\sin \alpha}.$$

In this region $\beta \ge 30°$ is equivalent to $\cot \beta \le \sqrt{3}$, hence it is the same as

$$\frac{a \perp \cos \alpha}{\sin \alpha} \le \sqrt{3},$$

or

$$a \le \cos \alpha + \sqrt{3} \sin \alpha.$$

This is identical to the statement we wanted to prove.

1967/2. *Prove that a tetrahedron with just one edge of length greater than* 1 *has volume at most* $\dfrac{1}{8}$.

Solution. The main idea of the solution can be summarized as follows: Let the longest edge of the tetrahedron $ABCD$ be CD, and the length of the opposite edge AB be $x \le 1$. We will give an estimate for the volume of the tetrahedron as a function of x.

Let us first estimate the altitude CT of the triangle ABC. Suppose that on AB, A is the closest vertex to T. Then

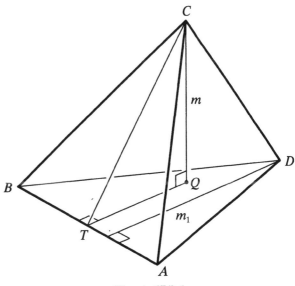

Figure 67/2.1

$BT \geq \dfrac{x}{2}$, and from the right triangle BTC we conclude (*Figure 1967/2.1*):

(1) $$CT = \sqrt{BC^2 - BT^2} \leq \sqrt{1 - \frac{x^2}{4}}.$$

Similar reasoning gives that the altitude m_1 of ABD belonging to AB satisfies

(2) $$m_1 \leq \sqrt{1 - \frac{x^2}{4}}.$$

The altitude $CQ = m$ of the tetrahedron cannot be more than CT, therefore

(3) $$m \leq \sqrt{1 - \frac{x^2}{4}}.$$

Now for the volume V of the tetrahedron the expression $V = \dfrac{1}{3} \cdot \dfrac{1}{2} AB \cdot m_1 \cdot$ $\cdot\, m$ together with (1), (2) and (3) implies

(4) $$V \leq \frac{x}{6}\left(1 - \frac{x^2}{4}\right) = \frac{1}{24} x(4 - x^2).$$

We have to show that $V \leq \dfrac{1}{8}$, which means $x(4 - x^2) \leq 3$ according to (4). This is true, because

$$3 - x(4 - x^2) = (1 - x)(3 - x - x^2) \geq 0,$$

since in $[0, 1]$ we have $1 - x \geq 0$ and $3 - x - x^2 \geq 0$.

Remarks. 1. The function $x(4 - x^2)$ in $[0, 1]$ considers its maximum at $x = 1$, and in a tetrahedron corresponding to this maximum the faces ABC and ABD are equilateral triangles with sides of unit length, and their planes are orthogonal to each other.

2. The maxima of the function under (4) cannot be directly determined using differentiation since the zeros of the derivative are outside of the unit interval $[0, 1]$. In fact, the derivative $\dfrac{1}{24}(4 - 3x^2)$ is positive and monotone increasing in $[0, 1]$ hence it takes its maximum at the endpoint $x = 1$.

1967/3. *Let k, m, n be positive integers such that $m + k + 1$ is a prime greater than $(n + 1)$ and let $c_s = s(s + 1)$ ($s = 1, 2, \ldots$). Prove that the product*

(1) $$(c_{m+1} - c_k)(c_{m+2} - c_k) \cdot \ldots \cdot (c_{m+n} - c_k)$$

is divisible by

(2) $$c_1 c_2 \ldots c_n.$$

Solution. Let P and Q denote the products under (1) and (2). First we will examine these expressions in detail. A generic term in P admits the form

$$c_a - c_b = a(a + 1) - b(b + 1) = a^2 - b^2 + a - b = (a - b)(a + b + 1).$$

From this we get that

$$P = (m \perp k + 1)(m + k + 2)(m \perp k + 2)(m + k + 3) \ldots (m \perp k + n)(m + k + n + 1) =$$
$$= ((m{\perp}k{+}1)(m{\perp}k{+}2) \ldots (m{\perp}k{+}n))((m{+}k{+}2)(m{+}k{+}3) \ldots (m{+}k{+}n{+}1)).$$

Let us introduce the following notations:

$$A = (m \perp k + 1)(m \perp k + 2) \ldots (m \perp k + n),$$
$$B = (m + k + 2)(m + k + 3) \ldots (m + k + n + 1).$$

With these notations $P = AB$. Next we focus on Q:

$$Q = 1(1 + 1)2(2 + 1) \ldots n(n + 1) = (1 \cdot 2 \cdot \ldots \cdot n)(2 \cdot 3 \cdot \ldots \cdot (n + 1)) = n!(n + 1)!$$

In conclusion, we have to prove that

$$\frac{AB}{n!(n+1)!} = \frac{A}{n!} \cdot \frac{B}{(n+1)!} \quad \text{is an integer.}$$

To achieve this, it is enough to confirm that $\dfrac{A}{n!}$ and $\dfrac{B}{(n+1)!}$ are both integers.

Now $\dfrac{A}{n!}$ is an integer, since by its definition A is the product of n consecutive integers, hence it is always divisible by $n!$ (see also the remark following the solution).

In order to show that $\dfrac{B}{(n+1)!}$ is also an integer, notice that it is an integer if and only if $\dfrac{(m+k+1)B}{(n+1)!}$ is an integer, since (according to the problem) $m + k + 1$ is a prime larger than $n + 1$. According to the definition of B the expression $(m + k + 1)B$ is the product of $n + 1$ consecutive integers, so it is divisible by $(n + 1)!$. This last argument now completes our solution.

Remark. In our solution we used the fact that the product of n consecutive integers is divisible by $n!$. It means that if the first term is $k + 1$ then

$$E = \frac{(k + 1)(k + 2) \ldots (k + n)}{n!}$$

is an integer. Notice that

$$E = \frac{k!(k + 1) \ldots (k + n)}{k! n!} = \frac{(n + k)!}{k! n!} = \binom{n + k}{k},$$

which (according to basic properties of binomial coefficients) is an integer.

1967/4. *$A_0 B_0 C_0$ and $A'B'C'$ are given acute triangles. Construct the triangle ABC with the largest possible area which is circumscribed around $A_0 B_0 C_0$ (i.e., AB contains C_0, BC contains A_0 and CA contains B_0) and is similar to $A'B'C'$ (A, B, C correspond to A', B' and C').*

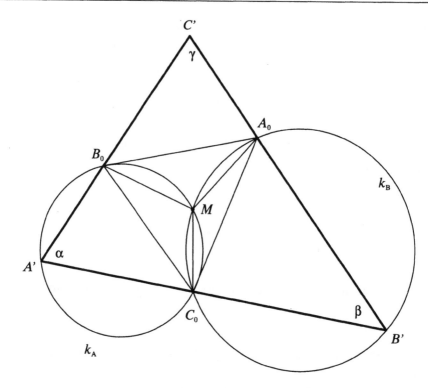

Figure 67/4.1

Solution. First let us inscribe a triangle in $A'B'C'$ similar to $A_0B_0C_0$ in such a way that its vertices are on the edges $B'C'$, $C'A'$ and $A'B'$. This is possible, since for an arbitrary interval $\overline{C_0}\overline{B_0}$ parallel to $B'C'$ a triangle similar to $A_0B_0C_0$ can be constructed on its side opposite to A'. Then it can be shrunk from A' until the vertex corresponding to A_0 will lie on $B'C'$. For the sake of simplicity this inscribed triangle will be denoted by $A_0B_0C_0$ as well (*Figure 1967/4.1*).

We denote the intersection (other than C_0) of the circumcircles k_A and k_B of $A'B_0C_0$ and $B'A_0C_0$ by M. Since $A'B'C'$ and $A_0B_0C_0$ are both acute, M is an inner point of the triangle $A_0B_0C_0$ and hence of $A'B'C'$. It is on the arcs B_0C_0 and A_0C_0 of the circles k_A and k_B not containing A' and B'. Therefore if the angles of $A'B'C'$ are α, β, γ, then

$$\angle B_0MC_0 = 180° \perp \alpha, \qquad \angle A_0MC_0 = 180° \perp \beta,$$

and so

$$\angle A_0MB_0 = 360° \perp (180° \perp \alpha) \perp (180° \perp \beta) = \alpha + \beta = 180° \perp \gamma.$$

This shows that M is on the arc A_0B_0 of the circumcircle of $C'B_0A_0$ not containing C'.

Since the rectangles $MB_0A'C_0$, $MC_0B'A_0$ and $MA_0C'B_0$ admit circumcircles, we have $\angle A'C_0M = \angle B'A_0M = \angle C'B_0M$ (*Figure 1967/4.2*). Let φ_1 and

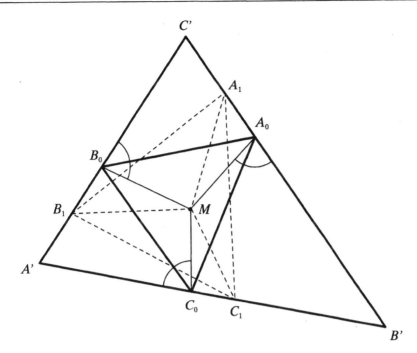

Figure 67/4.2

φ_2 denote stretching-rotations with centre M taking the line $A'B'$ into $B'C'$ and $B'C'$ into $C'A'$. It follows that $\varphi_1(C_0) = A_0$ and $\varphi_2(A_0) = B_0$. For an arbitrary point C_1 on $A'B'$ with $\varphi_1(C_1) = A_1$ and $\varphi_2(A_1) = B_1$, the triangles $A_0B_0C_0$ and $A_1B_1C_1$ are similar since they are composed from the similar triangles $C_0MA_0 \sim C_1MA_1$, $A_0MB_0 \sim A_1MB_1$. Therefore starting from a point C_1 on $A'B'$, infinitely many triangles inscribed in $A'B'C'$ and similar to $A_0B_0C_0$ can be constructed.

Next we will show that M is determined by the triangles $A'B'C'$ and $A_0B_0C_0$: For another M^* with the same properties we could achieve that sides of $\overline{A_0}\overline{B_0}\overline{C_0}$ were parallel to corresponding sides of $A_0B_0C_0$, and so in $A'B'C'$ there are two similar triangles in similar positions which is clearly a contradiction.

In conclusion, with the aid of the point M we can describe all triangles inscribed in $A'B'C'$ in the required way. The one with the minimal area will obviously have the feet of the orthogonals from M to the sides as vertices. These feet are interior points of the sides since $A'B'C'$ is acute and M is an inner point.

The construction concludes by shrinking the triangle we got until the resulting $A_0B_0C_0$ will become congruent to the triangle given by the problem. If $A_0B_0C_0$ was the inscribed triangle with the smallest possible area then the corresponding circumscribed triangle has the largest possible area.

Remark. Of course the problem has many other solutions. The difficulty in evaluating such a solution is to find the appropriate amount of precisity when discussing facts which are obvious when visualized. The solution given above roughly reflects how much is expected in the competition. For example, the fact that M is in $A_0 B_0 C_0$ is fairly obvious, but a rigorous proof requires quite an amount of work.

1967/5. *Consider the sequence $\{c_n\}$ given as*

$$c_1 = a_1 + a_2 + \ldots + a_8$$
$$c_2 = a_1^2 + a_2^2 + \ldots + a_8^2$$

$$\vdots$$

$$c_n = a_1^n + a_2^n + \ldots + a_8^n,$$

$$\vdots$$

for $a_1, a_2, \ldots, a_8$ reals, not all zero.

Given that an infinite number of $\{c_n\}$ is zero, find all n for which $c_n = 0$.

Solution. If n is even, then the summands defining it are all nonnegative and contain nonzero terms, hence $c_n \neq 0$.

Suppose now that n is odd. There are positive and negative reals among a_i otherwise the sum of their (odd) powers could not be zero. Let a_1 has the largest absolute value, while a_2 has the largest absolute value among those having opposite sign than a_1. Obviously, a_1 and a_2 are nonzero. Write c_n as

(1) $$c_n = a_1^n \left(1 + \left(\frac{a_2}{a_1} \right)^n + \left(\frac{a_3}{a_1} \right)^n + \ldots + \left(\frac{a_8}{a_1} \right)^n \right) = a_1^n S.$$

The term $\dfrac{a_2}{a_1}$ is negative and not more than any other $\dfrac{a_i}{a_1}$, and the same holds for all their odd nth powers. Therefore

(2) $$S \geq 1 + 7 \left(\frac{a_2}{a_1} \right)^n.$$

According to (1) now c_n is zero only in case $S = 0$ since $a_1 \neq 0$. If $|a_1| \neq |a_2|$ (i.e. $|a_2| < |a_1|$) then $\left(\dfrac{a_2}{a_1} \right)^n$ converges to zero, hence after some n_0 the expression $\left| \dfrac{a_2}{a_1} \right|^n$ is less than $\dfrac{1}{7}$. In that case the right hand side of (2) is positive; and the same holds for S and c_n. This implies that c_n is 0 only in finitely many cases, showing that $|a_1| \neq |a_2|$ is impossible, so $a_2 = \perp a_1$, in other words $a_1 + a_2 = 0$. This means that $a_1^n + a_2^n = 0$ for all odd n.

If all $a_3, a_4, \ldots, a_8$ are zero, then we are done. If there are nonzero reals among these six numbers then the method given above can be applied again. In conclusion we get that the a_i's can be paired up to give 0 summands for odd powers, hence $c_n = 0$ for all odd n and these are the only indices with $c_n = 0$.

1967/6. *In a sports contest a total of m medals were awarded over $n > 1$ days. On the first day one medal and $\frac{1}{7}$ of the remaining medals were awarded. On the second day two medal and $\frac{1}{7}$ of the remaining medals were awarded, and so on. On the last day the remaining n medals were awarded. How many medals and over how many days were awarded?*

First solution. The number of medals awarded on the kth day will be denoted by e_k, the number of medals available at the beginning of the same day is m_k. According to the assumptions

$$(1) \qquad e_k = k + \frac{m_k - k}{7} = \frac{m_k}{7} + \frac{6k}{7}.$$

Using $m_{k-1} - m_k = e_{k-1}$ we would like to find some connection between e_k and e_{k-1}:

$$e_{k-1} - e_k = \frac{m_{k-1} - m_k}{7} - \frac{6}{7} = \frac{e_{k-1}}{7} - \frac{6}{7},$$

$$(2) \qquad e_{k-1} = \frac{7e_k}{6} - 1.$$

According to the problem $e_n = n$, therefore $e_{n-1} = \frac{7n - 6}{6}$. This shows that n is divisible by 6, so $n \geq 6$. With $p = n - 6$ ($p \geq 0$) we get

$$e_{n-1} = \frac{7e_n}{6} - 1 = \frac{7n}{6} - 1 = \frac{7p}{6} + 6,$$

so (2) implies

$$e_{n-2} = \left(\frac{7}{6}\right)^2 p + 6, \quad e_{n-3} = \left(\frac{7}{6}\right)^3 p + 6, \ldots, e_1 = \left(\frac{7}{6}\right)^{n-1} p + 6.$$

e_1 is an integer only if $p = n - 6$ is divisible by 6^{n-1}. This, however, implies $n = 6$ since $n - 6 < 6^{n-1}$. Therefore, the solution is:

$$n = 6, \quad e_1 = e_2 = \ldots = e_6 = 6 \quad \text{and} \quad m = 36.$$

Second solution. We will count the medals which remained after each day: Since $m_k - m_{k+1} = e_k$, (1) shows

$$m_{k+1} = m_k - e_k = m_k - k - \frac{m_k - k}{7} = \frac{6(m_k - k)}{7}.$$

For the sake of brevity let q denote $\frac{7}{6}$, and express m_k as

$$m_k = k + q m_{k+1}.$$

Repeating this process we get

$$m_n = n$$
$$m_{n-1} = (n-1) + qn,$$
$$m_{n-2} = (n-2) + q(n-1) + q^2 n,$$

$$\ldots$$

$$m = m_1 = 1 + 2q + 3q^2 + \ldots + nq^{n-1}.$$

Writing the sum on the right hand side in closed form we arrive to

(3) $\quad m = 1 + 2q + 3q^2 + \ldots + nq^{n-1} = \dfrac{nq^{n+1} - (n+1)q^n + 1}{(q-1)^2} = \dfrac{7^n(n-6)}{6^{n-1}} + 6^2.$

m is an integers only in case $n-6$ is divisible by 6^{n-1} and since $n-6 < 6^{n-1}$ this implies $n = 6$, showing that $e_1 = e_2 = \ldots = e_6 = 6$ and $m = 36$.

Remarks. 1. The proof of the identity in (3) goes as follows

$$S = 1 + 2q + 3q^2 + \ldots + \quad nq^{n-1}$$
$$Sq = \quad q + 2q^2 + 3q^3 + (n-1)q^{n-1} + nq^n$$
$$Sq - S = S(q-1) = -(1 + q + q^2 + \ldots + q^{n-1}) + nq^n =$$
$$= nq^n - \frac{q^n - 1}{q-1} = \frac{nq^{n+1} - (n+1)q^n + 1}{q-1},$$

implying

$$S = \frac{nq^{n+1} - (n+1)q^n + 1}{(q-1)^2}.$$

2. We just mention here that the Hungarian sports press dealt with the IMO only because of the above problem. A magazine blamed the organizers that by posing such "trivial" problems, the olympic spirit had been discredited. The members of the Hungarian team invited the editors to their annual meeting and asked them to show the solution of this trivial problem. Needless to say, the answer and the solution has not been arrived at in the past 25 years ...

1968.

1968/1. *Show that there is a unique triangle whose side lengths are consecutive integers and one of whose angles is twice another.*

First solution. With the usual notations let $\beta = 2\alpha$. The bisector at B intersects AC in B' (*Figure 1968/1.1*) and divides it into subintervals of lengths $CB' = \dfrac{ab}{a+c}$ and $B'A = \dfrac{cb}{a+c}$. (This is because an angle bisector divides the opposite side with the ratio of the adjacent sides.) Since their angles agree, ABC

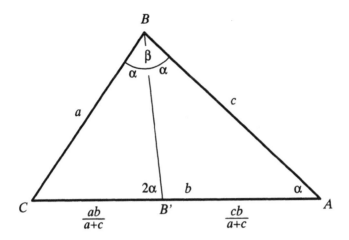

Figure 68/1.1

and $BB'C$ are similar triangles. Therefore

(1)
$$\frac{BC}{AC} = \frac{CB'}{BC}, \quad \text{i.e. } BC^2 = AC \cdot CB',$$

(2)
$$a^2 = \frac{ab^2}{a+c}, \quad b^2 = a(a+c).$$

According to our assumption on the angles $b > a$ holds; this implies that either $b = a+1$ or $b = a+2$.

In the first case

$$(a+1)^2 = a(a+c), \qquad c = 2 + \frac{1}{a},$$

and since c is an integer, this shows $a = 1$, and so $b = 2$ and $c = 3$; there is, however, no triangle with these sides.

In the second case $b = a+2$, but since a, b, c are consecutive integers, c is between a and b: $c = a+1$. Now (2) gives

$$(a+2)^2 = a(2a+1), \quad a^2 \perp 3a \perp 4 = 0,$$

$$(a+1)(a \perp 4) = 0.$$

The unique positive solution of this equation is $a = 4$, implying $b = 6$ and $c = 5$. In conclusion there is at most one triangle satisfying the assumption of the problem. In order to show that in the triangle of sides 4, 5 and 6 the angle opposite to the side of length 6 is twice the one opposite to the side of length 4, we denote the edges as follows: $AB = 5$, $BC = 4$, $CA = 6$. As we noted towards the beginning of the solution, we have $CB' = \frac{24}{9}$. The triangles ABC and $BB'C$ are similar (they share an angle at C and the ratio of the adjacent sides). This implies that $\angle CBB' = \angle CAB = \alpha$, hence the angle at B is 2α.

Second solution. The key ingredient of the above solution is formula (2) — this formula can be derived using trigonometry as follows: Starting with $\beta = 2\alpha$ we apply the law of sines:

$$\frac{b}{a} = \frac{\sin 2\alpha}{\sin \alpha} = 2\cos\alpha = \frac{b^2 + c^2 - a^2}{bc},$$

$$b^2 c = ab^2 + ac^2 - a^3, \qquad b^2(c - a) = a(a + c)(c - a).$$

Since $c \neq a$, it shows

$$b^2 = a(a + c).$$

Now we complete the solution as above.

1968/2. *Find all positive integers x for which*

(1) $$p(x) = x^2 - 10x - 22,$$

where $p(x)$ is the product of the decimal digits of x.

First solution. After taking a few examples it is not hard to see that the number of digits of x should not exceed 2: if n is the number of digits of x and $n > 2$, then — since the product of the digits is at most 9^n:

$$9^n \geq p(x) = x(x - 10) - 22 \geq 10^{n-1}(10^{n-1} - 10) - 22 =$$
$$= 10^n(10^{n-2} - 1) - 22 > 10^n,$$

providing a contradiction. Since $p(x) = x^2 - 10x - 22 = x$ does not admit integer solution, the number of digits cannot be one either. Thus we get that

$$x = 10a + b,$$

where $1 \leq a \leq 9$, $0 \leq b \leq 9$. It transforms the equation into

$$(10a + b)^2 - 10(10a + b) - 22 = ab,$$

which is equivalent to

(2) $$100a(a - 1) + (19a - 10)b + (b^2 - 22) = 0.$$

If $a \geq 2$, the left hand side is at least

$$200 + 28b + b^2 - 22 = b^2 + 28b + 178,$$

which is strictly positive. Therefore $a = 1$, and now (2) shows

$$b^2 + 9b - 22 = 0.$$

The positive solution of this equation is 2, hence $x = 12$, and this number satisfies (1).

Second solution. $p(x)=0$ does not admit integer solution, hence $p(x)$ is positive, so
$$x^2 - 10x - 22 > 0, \qquad (x>0)$$
implying

(3) $$x > 11{,}86.$$

If x is of n digits with first digit equal to A then $x = 10^{n-1}A + B$, with B of $n-1$ digits. From this the maximal value of the product of digits of x is $A \cdot 9^{n-1}$, hence
$$p(x) \le A \cdot 9^{n-1} < 10^{n-1}A \le 10^{n-1}A + B = x$$
so
$$x^2 - 11x - 22 < 0,$$
implying

(4) $$x < 12{,}73.$$

Based on (3) and (4) we get that $11{,}86 < x < 12{,}73$, and since x is an integer, the only possible solution is $x = 12$, which number eventually solves the problem.

1968/3. *Prove that the system*
$$ax_1^2 + bx_1 + c = x_2,$$
$$ax_2^2 + bx_2 + c = x_3,$$

(1)

$$\cdots\cdots\cdots\cdots$$
$$ax_n^2 + bx_n + c = x_1$$

(a, b, c are real with $a \neq 0$)

  I. *has no real solution if* $(b-1)^2 - 4ac < 0$;

 II. *has one real solution if* $(b-1)^2 - 4ac = 0$;

III. *and has more than one real solution once* $(b-1)^2 - 4ac > 0$.

Solution. Notice that $(b-1)^2 - 4ac$ is equal to the discriminant of the quadratic equation

(2) $$f(x) = ax^2 + (b-1)x + c = 0.$$

The ith equation in (1) can be written as
$$ax_i^2 + bx_i + c = x_{i+1} \qquad (n+1 := 1)$$

(3) $$f(x_i) = ax_i^2 + bx_i - x_i + c = x_{i+1} - x_i;$$

adding these expressions for $i = 1, 2, \ldots, n$ we get

(4) $$f(x_1) + f(x_2) + \ldots + f(x_n) = 0.$$

Hence for a solution $x_1, x_2, \ldots, x_n$ of (1) the above equality is satisfied.

  I. In case $(b-1)^2 - 4ac < 0$ the polynomial $f(x)$ has no real solution, hence it is either positive or negative for all x, therefore the left hand side of (4) cannot be 0. This shows that (1) admits no solution in this case.

II. If $(b \perp 1)^2 \perp 4ac = 0$ then $f(x)$ is either nonnegative or nonpositive for all x. Therefore (4) holds only if

$$f(x_1) = f(x_2) = \ldots = f(x_4) = 0.$$

The unique solution of $f(x) = 0$ is $x = \dfrac{1 \perp b}{2a}$, hence the unique solution of (1) is

$$x_1 = x_2 = \ldots = x_n = \frac{1 \perp b}{2a}.$$

III. Finally if $(b \perp 1)^2 \perp 4ac > 0$, then $f(x)$ has two different real roots y_1 and y_2. Now

$$x_1 = x_2 = \ldots = x_n = y_1, \quad \text{and} \quad x_1 = x_2 = \ldots = x_n = y_2$$

both satisfy (1), hence the system admits more than one solution.

Remark. Notice that III. does not prove that in that case the system admits *exactly* two solutions; we only showed that there are *at least* two different solutions. In fact, in certain cases there are more solutions, for example

$$x_1^2 \perp 4{,}5x_1 + 6 = x_2$$
$$x_2^2 \perp 4{,}5x_2 + 6 = x_1$$

is solved by the following four pairs:

$$(1{,}5; 1{,}5) \quad (4; 4), \quad (1; 2{,}5) \quad (2{,}5; 1).$$

1968/4. *Prove that every tetrahedron has a vertex whose three edges have the right lengths to form a triangle.*

Solution. The intervals a, b, c form a triangle if and only if the following (so-called triangle) inequalities are satisfied:

$$a < b+c, \quad \text{i.e.} \quad \perp a+b+c > 0,$$
$$b < c+a, \quad \text{i.e.} \quad a \perp b+c > 0,$$
$$c < a+b, \quad \text{i.e.} \quad a+b \perp c > 0.$$

If a is the longest among the three intervals then it is enough to check the first inequality, the other two are automatically satisfied.

Label the vertices of the tetrahedron in a way that the longest (or one of the longest) edge is be denoted by AB. Apply the triangle inequality for the triangles ABC and ABD:

$$\perp AB + BC + AC > 0,$$
$$\perp AB + BD + AD > 0.$$

Adding up these inequalities we get

$$(\perp AB + AC + AD) + (\perp AB + BC + BD) > 0.$$

This means that at least one expression in the parentheses must be nonnegative.

If $\angle AB + AC + AD > 0$ then AB, AC and AD form a triangle, since AB is (by our assumption) the longest edge in the tetrahedron.

In case $\angle AB + BC + BD > 0$, the edges AB, BC, BD form a triangle. In conclusion the edges of A or B form a triangle.

Remark. For tetrahedrons with the property that the two longest edges are opposite, one can find a vertex whose edges form an acute triangle.

1968/5. *Let f be a real-valued function defined for all real numbers, such that for some $a > 0$ it satisfies*

(1)
$$f(x+a) = \frac{1}{2} + \sqrt{f(x) - (f(x))^2}.$$

I. *Prove that f is periodic, i.e., there exists a positive real b such that*

$$f(x+b) = f(x)$$

holds for every x.

II. *Give an example of such a non-constant f for $a = 1$.*

Before proceeding further we note that

(2)
$$\frac{1}{2} \leq f(x) \leq 1$$

holds for all x since $f(x)$ is the sum of $\frac{1}{2}$ and a nonnegative real, and the expression under the square root is nonnegative only if

$$f(x)(1 - f(x)) \geq 0,$$

which (according to $f(x) \geq \frac{1}{2}$) implies $f(x) \leq 1$.

First solution. I. Since (1) provides a relation between $f(x+a)$ and $f(x)$, it seems plausible to examine the substitution $x \to x + a$ in (1):

$$f(x+2a) = \frac{1}{2} + \sqrt{f(x+a) - (f(x+a))^2} =$$

$$= \frac{1}{2} + \sqrt{\frac{1}{2} + \sqrt{f(x) - (f(x))^2} - \left(\frac{1}{2} - \sqrt{f(x) - (f(x))^2}\right)^2} =$$

$$= \frac{1}{2} + \sqrt{\frac{1}{2} + \sqrt{f(x) - (f(x))^2} - \frac{1}{4} - f(x) + (f(x))^2 - \sqrt{f(x) - (f(x))^2}} =$$

$$= \frac{1}{2} + \sqrt{\frac{1}{4} - f(x) + (f(x))^2} = \frac{1}{2} + \sqrt{\left(f(x) - \frac{1}{2}\right)^2} = \frac{1}{2} + f(x) - \frac{1}{2} = f(x).$$

This exactly shows that $b = 2a$ is a period of f.

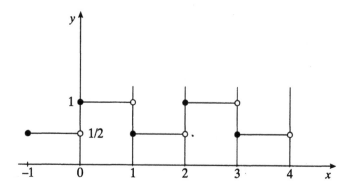

Figure 68/5.1

II. A simple example satisfying the assumptions of the problem can be given by the following formula (see also *Figure 1968/5.1*):

$$f(x)=\begin{cases}1, & \text{if } 2n\le x<2n+1,\\ \dfrac{1}{2} & \text{if } 2n+1\le x<2n+2,\end{cases} \quad n \text{ arbitrary integer.}$$

This function obviously satisfies (1).

Second solution. Regard (1) as a quadratic equation with $f(x)$ as the unknown. Taking its square we get

$$f(x+a)\perp\frac{1}{2}=\sqrt{f(x)\perp(f(x))^2}$$

$$f(x)^2\perp f(x)+\left(f(x+a)\perp\frac{1}{2}\right)^2=0$$

The solutions are provided by the quadratic formula:

$$\frac{1}{2}+\sqrt{f(x+a)\perp(f(x+a))^2}, \qquad \frac{1}{2}\perp\sqrt{f(x+a)\perp(f(x+a))^2}.$$

The second root is less than $\dfrac{1}{2}$, hence the unique solution for $f(x)$ is:

$$f(x)=\frac{1}{2}+\sqrt{f(x+a)\perp(f(x+a))^2}.$$

Now perform the substitution $x\to x\perp a$ and get

$$f(x\perp a)=\frac{1}{2}+\sqrt{f(x)\perp(f(x))^2}.$$

According to (1) this, however, shows that

$$f(x\perp a)=f(x+a),$$

hence a final substitution $x\to x+a$ implies

$$f(x)=f(x+2a),$$

consequently $b=2a$ is a period of f.

Remark. There are many ways to produce an example of a function with the required properties. A continuous function with period 2 ($a = 1$) can be given by the formula

$$f(x) = \frac{1}{2}\left(1 + \left|\sin\frac{\pi x}{2}\right|\right).$$

1968/6. *Let* $[x]$ *denote the greatest integer not larger than* x *(the "integer part" of* x*). For every positive integer* n *evaluate the sum*

(1)
$$\left[\frac{n+1}{2}\right] + \left[\frac{n+2}{2^2}\right] + \ldots + \left[\frac{n+2^k}{2^{k+1}}\right] + \ldots$$

First solution. Notice first that the sum in (1) is finite, since once $2^k > n$ we have $n < 2^k = 2^k(2 \perp 1) = 2^{k+1} \perp 2^k$, and so

$$\frac{n+2^k}{2^{k+1}} < 1, \quad \text{therefore} \quad \left[\frac{n+2^k}{2^{k+1}}\right] = 0.$$

First we prove the following statement: for any real x

(2)
$$\left[x + \frac{1}{2}\right] = [2x] \perp [x].$$

Any real x can be written either as $m + \alpha$ or as $m + \frac{1}{2} + \alpha$, where m is an integer and $0 \leq \alpha < \frac{1}{2}$. In the first case

$$[x] = m, \quad \left[x + \frac{1}{2}\right] = m, \quad [2x] = [2m + 2\alpha] = 2m,$$

implying (2). In the second case

$$[x] = m, \quad \left[x + \frac{1}{2}\right] = m + 1, \quad [2x] = [2m + 1 + 2\alpha] = 2m + 1,$$

which again implies (2).

Using (2) (and assuming that $2^k > n$) now write (1) as:

$$\left[\frac{n+1}{2}\right] + \left[\frac{n+2}{2^2}\right] + \ldots + \left[\frac{n+2^k}{2^{k+1}}\right] + \ldots =$$

$$= \left[\frac{n}{2} + \frac{1}{2}\right] + \left[\frac{n}{2^2} + \frac{1}{2}\right] + \ldots + \left[\frac{n}{2^{k+1}} + \frac{1}{2}\right] + \ldots =$$

$$= [n] \perp \left[\frac{n}{2}\right] + \left[\frac{n}{2}\right] \perp \left[\frac{n}{2^2}\right] + \ldots + \left[\frac{n}{2^k}\right] \perp \left[\frac{n}{2^{k+1}}\right] = [n] \perp \left[\frac{n}{2^{k+1}}\right] = [n] = n,$$

since $\left[\frac{n}{2^{k+1}}\right] = 0$. Therefore the value of the sum in (1) is equal to n.

Second solution. Write n in base two:

$$n = a_s \cdot 2^s + a_{s-1} \cdot 2^{s-1} + \ldots + a_k \cdot 2^k + \ldots + a_1 \cdot 2 + a_0,$$

where $a_i = 0$ or 1 and $a_s = 1$. If $k > s$ then $k \geq s+1$, $2^k \geq 2^{s+1} > n$, and so

$$\frac{n+2^k}{2^{k+1}} < \frac{2^k + 2^k}{2^{k+1}} = 1, \qquad \left[\frac{n+2^k}{2^{k+1}}\right] = 0.$$

Therefore we can assume that $k \leq s$, hence

$$n + 2^k = a_s \cdot 2^s + a_{s-1} \cdot 2^{s-1} + \ldots + (a_k + 1)2^k + \ldots + a_1 \cdot 2 + a_0,$$

(3) $\qquad \dfrac{n+2^k}{2^{k+1}} = \left(a_s \cdot 2^{s-k-1} + a_{s-1} \cdot 2^{s-k-2} + \ldots + a_{k+1}\right) + \left((a_k + 1)2^{-1}\right) +$

$$+ \left(a_{k-1} \cdot 2^{-2} + \ldots + a_1 \cdot 2^{-k} + a_0 \cdot 2^{-k-1}\right).$$

The value of the first expression in (3) is an integer, while the value of the second one is 1 if $a_k = 1$ and $\dfrac{1}{2}$ if $a_k = 0$. The third term can be estimated as

$$a_{k-1}2^{-2} + a_{k-2}2^{-3} + \ldots + a_1 \cdot 2^{-k} + a_0 \cdot 2^{-k-1} \leq 2^{-2} + 2^{-3} + \ldots + 2^{-k-1} <$$

$$< 2^{-2} + 2^{-3} + \ldots < \frac{2^{-2}}{\frac{1}{2}} = \frac{1}{2}.$$

Therefore

$$\left[\frac{n+2^k}{2^{k+1}}\right] = a_s \cdot 2^{s-k-1} + a_{s-1} \cdot 2^{s-k-2} + a_{k+1} + a_k.$$

Consider this expression for all k between $k = 0$ and $k = s$:

$$\left[\frac{n+1}{2}\right] = a_s \cdot 2^{s-1} + a_{s-1} \cdot 2^{s-2} + \ldots + a_2 \cdot 2 + a_1 + a_0$$

$$\left[\frac{n+2}{4}\right] = a_s \cdot 2^{s-2} + a_{s-1} \cdot 2^{s-3} + \ldots + a_2 + a_1$$

$$\left[\frac{n+2^{s-1}}{2^s}\right] = a_s + a_{s-1}$$

$$\left[\frac{n+2^s}{2^{s+1}}\right] = a_s.$$

Summing these equalities we get that (1) is equal to

$$a_s \cdot 2^s + a_{s-1}2^{s-1} + \ldots + 2a_1 + a_0 = n.$$

Third solution. First we prove a useful formula: The cardinality of the set of those integers in $H = \{1, 2, \ldots, n\}$ which are divisible by 2^k but not by 2^{k+1}

is precisely

(4)
$$\left[\frac{n+2^k}{2^{k+1}}\right].$$

(These are the integers for which their prime decomposition contains 2 on power k. Partition the elements of H into H_0, H_1, ..., H_k in such a way that H_i contains those elements of H for which their prime decomposition contains 2 on power i. Obviously, the H_i's provide a cover of H with disjoint subsets.) The proof of the formula is quite easy: The number of elements in H divisible by 2^k is exactly $\left[\dfrac{n}{2^k}\right]$ and the number of elements divisible by 2^{k+1} is $\left[\dfrac{n}{2^{k+1}}\right]$. This implies that the number of elements of H divisible 2^k but not by 2^{k+1} is

$$\left[\frac{n}{2^k}\right] \perp \left[\frac{n}{2^{k+1}}\right].$$

Applying (2) for $x = \dfrac{n}{2^{k+1}}$ we get

$$\left[\frac{n}{2^k}\right] \perp \left[\frac{n}{2^{k+1}}\right] = \left[\frac{n}{2^{k+1}} + \frac{1}{2}\right] = \left[\frac{n+2^k}{2^{k+1}}\right],$$

which proves the statement in (4). Since the disjoint union of the H_i's is the set H containing n elements, and there are $\left[\dfrac{n+2^i}{2^{i+1}}\right]$ integers in H_i, we get that

$$\left[\frac{n+1}{2}\right] + \left[\frac{n+2}{2^2}\right] + \ldots + \left[\frac{n+2^k}{2^{k+1}}\right] = n.$$

Remark. The first solution applies even for the case of an arbitrary real n; in this case the value of the sum in (1) is equal to $[n]$.

1969.

1969/1. *Prove that there are infinitely many positive integers a such that*

$$z = n^4 + a$$

is not a prime for any positive integer n.

Solution. Let $a = 4b^4$ where $b > 1$ is an arbitrary integer. We show that z is not prime:

$$z = n^4 + 4b^4 + 4n^2b^2 \perp 4n^2b^2 = (n^2 + 2b^2)^2 \perp (2nb)^2 =$$
$$= (n^2 + 2b^2 + 2nb)(n^2 + 2b^2 \perp 2nb) = \left((n+b)^2 + b^2\right)\left((n \perp b)^2 + b^2\right).$$

Since $b > 1$, both factors are greater than 1, hence z is not prime.

1969/2. *Let*

(1) $$f(x)=\cos(a_1+x)+\frac{\cos(a_2+x)}{2}+\frac{\cos(a_3+x)}{2^2}+\ldots+\frac{\cos(a_n+x)}{2^{n-1}},$$

where $a_1, a_2, \ldots, a_n$ are real constants and x is a real variable. Prove that if $f(x_1)=f(x_2)=0$ then $x_2\perp x_1=m\pi$ for some integer m.

First solution. According to a trigonometric identity we have

$$\cos(a_i+x)=\cos a_i \cos x \perp \sin a_i \sin x;$$

applying this to the left hand side of (1) we get

$$f(x)=\left(\cos a_1+\frac{\cos a_2}{2}+\ldots+\frac{\cos a_n}{2^{n-1}}\right)\cos x\perp$$

$$\perp\left(\sin a_1+\frac{\sin a_2}{2}+\ldots+\frac{\sin a_n}{2^{n-1}}\right)\sin x=A\cos x\perp B\sin x.$$

A and B cannot simultaneously vanish, since this would imply $f(x)=0$ for all x, although

(2) $$f(\perp a_1)=1+\frac{\cos(a_2\perp a_1)}{2}+\ldots+\frac{\cos(a_n\perp a_1)}{2^{n-1}}\geq$$

$$\geq 1\perp\left(\frac{1}{2}+\frac{1}{2^2}+\ldots+\frac{1}{2^{n-1}}\right)=\frac{1}{2^{n-1}}>0$$

(since $\cos(a_i\perp a_1)\geq\perp 1$). Let us write the above expression for $f(x)$ in the following form:

$$f(x)=A\cos x\perp B\sin x=\sqrt{A^2+B^2}\left(\frac{A}{\sqrt{A^2+B^2}}\cos x\perp\frac{B}{\sqrt{A^2+B^2}}\sin x\right).$$

Let a real number φ be chosen in such a way that $\cos\varphi=\dfrac{A}{\sqrt{A^2+B^2}}$ and so $\sin\varphi=\dfrac{B}{\sqrt{A^2+B^2}}$ hold. With this notation at hand

$$f(x)=\sqrt{A^2+B^2}(\cos x\cos\varphi\perp\sin x\sin\varphi)=\sqrt{A^2+B^2}\cos(x+\varphi).$$

Since the distance between two roots of cosine is an integer multiple of π, for

$$f(x_1)=f(x_2)=0$$

we get

$$\cos(x_1+\varphi)=\cos(x_2+\varphi)=0,$$

therefore

$$(x_2+\varphi)\perp(x_1+\varphi)=x_2\perp x_1=m\pi.$$

The proof is now complete.

Second solution. Let z_k be the complex number with absolute value $\dfrac{1}{2^{k-1}}$ and argument a_k:

$$z_k = \frac{1}{2^{k-1}}(\cos a_k + i \sin a_k) \qquad (k = 1, 2, \ldots, n),$$

and let

$$z = \cos x + i \sin x.$$

The real part of a complex number c will be denoted by $\Re c$.

With these notations $z_k z = \dfrac{1}{2^{k-1}}\big(\cos(a_k + x) + 2\sin(a_k + x)\big)$, and so

(3) $$f(x) = \Re(z_1 z + z_2 z + \ldots + z_n z) = \Re\big(z(z_1 + z_2 + \ldots + z_n)\big).$$

It is not hard to see that $z_1 + z_2 + \ldots + z_n \neq 0$, since otherwise

$$z_1 = \bot(z_2 + z_3 + \ldots + z_n)$$
$$|z_1| = |z_2 + z_3 + \ldots + z_n| \leq |z_2| + |z_3| + \ldots + |z_n|$$

would imply

$$1 \leq \frac{1}{2} + \frac{1}{2^2} + \ldots + \frac{1}{2^{n-1}} = 1 \bot \frac{1}{2^{n-1}}.$$

This last inequality is clearly a contradiction, hence we get that $z_1 + z_2 + \ldots + z_n = c \neq 0$. Let $c = r(\cos \varphi + i \sin \varphi)$; now (3) implies

$$f(x) = \Re(cz) = r \cos(x + \varphi), \quad (r \neq 0).$$

If $f(x_1) = f(x_2) = 0$, then

$$\cos(x_1 + \varphi) = \cos(x_2 + \varphi) = 0,$$

showing that

$$x_2 + \varphi \bot (x_1 + \varphi) = x_2 \bot x_1 = m\pi \qquad (m \text{ integer}),$$

completing the proof.

1969/3. *For each $k = 1, 2, 3, 4, 5$ find necessary and sufficient conditions on $a > 0$ such that there exists a tetrahedron with k edges of length a and $(6 \bot k)$ edges of length 1.*

Solution. Let us examine the various values of k one after another.

A) $k = 1$.

For the tetrahedron $ABCD$ we have $AB = a$ and all the other edges are of unit length. The midpoint of CD is denoted by F (*Figure 1969/3.1*). From the equilateral triangles CDA and CDB we get $AF = BF = \dfrac{\sqrt{3}}{2}$, therefore the condition of the existence of the isosceles triangle ABF is:

(1) $$a < 2 \cdot \frac{\sqrt{3}}{2} = \sqrt{3}.$$

The above condition is also sufficient for the existence of the desired tetrahedron since with sides $AF = BF = \dfrac{\sqrt{3}}{2}$ and $AB = a$ a triangle ABF can be constructed

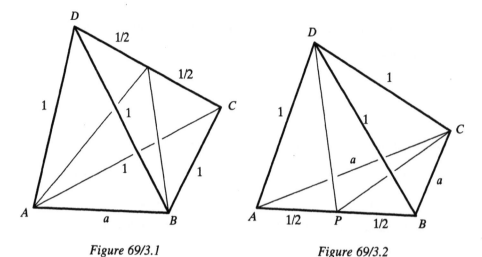

Figure 69/3.1 *Figure 69/3.2*

in the plane and by constructing an orthogonal to its plane of length $DF = FC = \frac{1}{2}$ at F we get a tetrahedron with edges of required lengths.

B) $k = 2$.

There are two possible choices for the two edges of length a. In case both of them are on the same face, say $AC = BC = a$ (*Figure 1969/3.2*) then consider the midpoint P of AB.

For the existence of ABC the inequality $2a > 1$, or equivalently $a > \frac{1}{2}$ is obviously necessary. Since $PD = \frac{\sqrt{3}}{2}$ and $PC = \sqrt{a^2 \perp \frac{1}{4}}$, the triangle inequality $PC + PD > CD$ for the triangle PCD yields

(2)
$$\sqrt{a^2 \perp \frac{1}{4}} + \frac{\sqrt{3}}{2} > 1.$$

Using (2) we get the necessary conditions

$$a^2 \perp \frac{1}{4} > \left(1 \perp \frac{\sqrt{3}}{2}\right)^2 = \frac{7}{4} \perp \sqrt{3}$$

and

(3)
$$a^2 > 2 \perp \sqrt{3}, \quad a > \sqrt{2 \perp \sqrt{3}}.$$

This condition is in accordance with $a > \frac{1}{2}$.

Similarly, from $PD + DC > PC$ we conclude

$$\frac{\sqrt{3}}{2} + 1 > \sqrt{a^2 \perp \frac{1}{4}},$$

and

(4) $$a^2 < 2+\sqrt{3}, \quad a < \sqrt{2+\sqrt{3}}.$$

The union of (3) and (4) now provides

(5) $$\sqrt{2 \perp \sqrt{3}} < a < \sqrt{2+\sqrt{3}}.$$

Since $CD + PC > PD$ automatically holds, in case (4) is satisfied, we can reverse our reasoning and this implies the existence of the triangle PCD. Now the orthogonals PA and PB in P of length $\frac{1}{2}$ and $\frac{1}{2}$ give the desired tetrahedron.

If the two edges of length a are skew, for example, $AB = CD = a$ (and all other edges are of unit length) then consider the midpoint Q of CD. As above, the triangle inequality for ABQ gives $AQ + BQ > AB$ (*Figure 1969/3.3*), and so

(6) $$2\sqrt{1 \perp \frac{a^2}{4}} > a, \quad \text{hence} \quad a < \sqrt{2}.$$

This last inequality is the necessary and sufficient condition for the existence of the triangle ABQ. Once ABQ is given, the orthogonals at Q with length $DQ = QC = \frac{a}{2}$ provide the tetrahedron. Hence in this case (6) is the condition we are looking for.

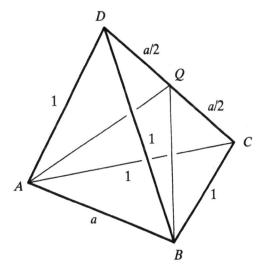

Figure 69/3.3

Summarizing the two cases above, if there are two edges of length a and four of length 1, then the necessary and sufficient condition for the existence of a tetrahedron with the given edges is that one of (5) and (6) has to be satisfied, which is equivalent to

(7) $$a < \sqrt{2+\sqrt{3}}.$$

C) $k = 3$.

We will show that for any choice of a there is such a tetrahedron. Choosing the edges of ABC to be equal to a and the remaining edges are of length 1, the distance of the centre of ABC from the vertices is $\dfrac{a\sqrt{3}}{3}$. This should be less than the sides,

(8)
$$\frac{a\sqrt{3}}{3} < 1, \quad a < \sqrt{3}.$$

Once this inequality is satisfied, a tetrahedron with base ABC and side edges of unit length exists.

If the base has edges of unit length, then (similarly to the above said) the necessary and sufficient condition is

(9)
$$\frac{1 \cdot \sqrt{3}}{3} < a, \quad a > \frac{1}{\sqrt{3}}.$$

Since one of (8) and (9) is always satisfied, we get that for $k = 3$ there is no restriction on the positive real number a.

D) $k = 4$.

This case is essentially the same as the case of $k = 2$; we just have to invert the roles of a and 1. This substitutes (7) with

$$1 < a\sqrt{2 + \sqrt{3}},$$

which gives

$$a > \sqrt{2 \perp \sqrt{3}}.$$

E) $k = 5$.

This case is basically the same as $k = 1$: again by changing the roles of a and 1 we get

$$1 < a\sqrt{3}, \quad a > \frac{1}{\sqrt{3}}.$$

In conclusion, the necessary and sufficient condition for the existence of a tetrahedron with the desired properties is:

$k = 1$:	$0 < a < \sqrt{3}$
$k = 2$:	$0 < a < \sqrt{2 + \sqrt{3}}$
$k = 3$:	$0 < a$
$k = 4$:	$\sqrt{2 \perp \sqrt{3}} < a$
$k = 5$:	$\dfrac{1}{\sqrt{3}} < a.$

1969/4. *Let C be an interior point of the semicircle k over AB and D is the foot of the perpendicular from C to AB. The circle k_1 is the incircle of ABC, the circle k_2 touches CD, DA and k while k_3 touches CD, DB and k. Show that k_1, k_2 and k_3 have another common tangent apart from AB.*

First solution. The centres and radii of k_1, k_2, k_3 will be denoted by O_1, O_2, O_3 and r_1, r_2, r_3, respectively. The tangent points of the circles on

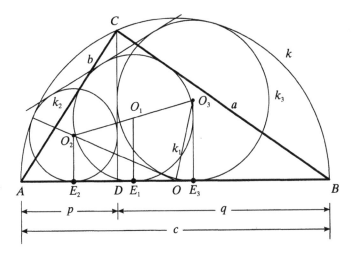

Figure 69/4.1

AB are E_1, E_2, E_3, where E_1 is in the interval E_2E_3 (*Figure 1969/4.1*). The statement of the problem is equivalent to saying that O_1, O_2, O_3 are collinear: The reflection of AB in this line provides the other common tangent.

We will use the following notations: $AB = c$, $BC = a$, $CA = b$, $AD = p$, $BD = q$, $a+b+c = 2s$. The circle k_1 divides the sides AB, BC, CA — as it is well-known — into intervals of lengths $s \perp a$, $s \perp b$; $s \perp b$, $s \perp c$; $s \perp c$, $s \perp a$ (*Figure 1969/4.2*). Since ABC is a right triangle we have $s \perp c = r_1$. The centre

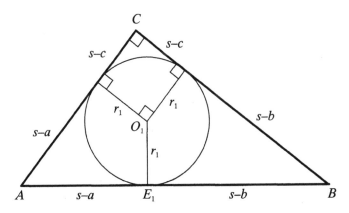

Figure 69/4.2

of k is O, and we assume that the notation is chosen such that $a \geq b$ is satisfied. From the right triangle O_2E_2O we get

$$\left(\frac{c}{2} \perp r_2^2\right)^2 = r_2^2 + \left(r_2 + q \perp \frac{c}{2}\right)^2,$$

$$cq = (r_2 + q)^2.$$

We also know that $cq = a^2$, therefore $a = r_2 + q$ and so

(1) $$r_2 = a \perp q.$$

Starting with $O_3 E_3 O$, a similar argument provides

(2) $$r_3 = b \perp p.$$

Now (1) and (2) implies

$$AE_2 = p \perp r_2 = p + q \perp a = c \perp a,$$
$$AE_3 = p + r_3 = b,$$

and

$$E_2 E_1 = AE_1 \perp AE_2 = (s \perp a) \perp (c \perp a) = s \perp c$$
$$E_1 E_3 = AE_3 \perp AE_1 = b \perp (s \perp a) = a + b \perp s = 2s \perp c \perp s = s \perp c.$$

This simply means that E_1 is the midpoint of $E_2 E_3$. Once again from (1) and (2) it follows that

$$r_2 + r_3 = a + b \perp (p + q) = a + b \perp c = 2(s \perp c) = 2r_1,$$

i.e., r_1 is equal to the arithmetic mean of r_2 and r_3. These last two results show that in the right trapezium $O_2 E_2 E_3 O_3$ the median starting at E_1 is of length r_1, hence its other endpoint O_1 is the midpoint of $O_2 O_3$. This observation now completes the solution.

Second solution. We will show that the statement of the problem holds for any triangle ABC with D an arbitrary inner point of AB and k the circumcircle of ABC.

In the solution we will use the fact proved in the first solution of Problem 1962/6., according to which if Q is the midpoint of the arc AB of the circumcircle k of ABC not containing C, then the centre K of the circumcircle is the point on the angle bisector CQ which satisfies $QA = QB = QK$.

Now let D be an arbitrary point of AB, and the circle k_2 touches AB in E, CD in F and the circumcircle k in T (*Figure 1969/4.3*).

T is the centre of similitude transforming k_2 into k; this enlargement maps E into Q, since in these points the tangents of k_2 and k are parallel. R on k corresponds to F, therefore the enlargement maps EF into QR, thus EF is parallel to QR. K will stand for the intersection of EF and CQ. We will show that K is the centre of the circumcircle of ABC, i.e. $K = O_1$; for this we have to verify that $QA = QK$.

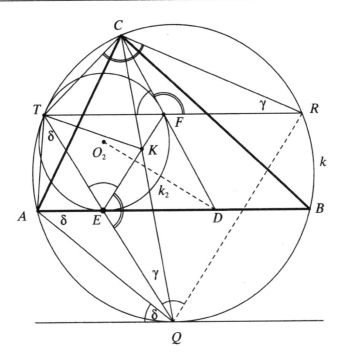

Figure 69/4.3

From the theorem regarding angles at the circumference it follows that

$$\angle TFC = \angle TEF = \angle TQR,$$

therefore their complements are equal as well; these are denoted by double arcs on the figure:

$$\angle CFR = \angle FEQ = \angle TCR.$$

Taking the equality $\angle TQK = \angle TRC = \gamma$ into account we conclude the similarity

(3) $$QEK \sim RFC \sim RCT$$

of the corresponding triangles. It follows from the intercept theorem that $\dfrac{QT}{RT} = \dfrac{QE}{RF}$, while the similarity of QEK and RFC yields $\dfrac{QE}{RF} = \dfrac{QK}{RC}$. Combining these two results we get

(4) $$\frac{QT}{RT} = \frac{QK}{RC}.$$

It implies the similarity of the triangles RCT and QKT since they share an angle (γ) and according to (4) the ratios of the edges adjacent to γ are equal.

(3) shows that $QEK \sim QKT$, hence

(5) $$\frac{QE}{QK} = \frac{QK}{QT}, \quad \text{consequently} \quad QK^2 = QE \cdot QT.$$

Furthermore, the triangles QEA and QAT are also similar since they share the angle $\angle AQE$, and $\angle QAE = \angle ATQ = \delta$ since these are both equal to the angle of AQ and the tangent at Q. From this similarity we deduce

$$\frac{QE}{QA} = \frac{QA}{QT}, \quad \text{i.e.} \quad QA^2 = QE \cdot QT,$$

and using (5) it implies $QA = QK$. Now this latter equality shows that K is the centre of the incircle k_1 of ABC; therefore the intersection of EF and CQ is K.

By repeating the above said for k_3 we get that the intersection of CQ with $E'F'$ is K; here E' and F' are the points of tangencies of k_3 with AB and CD, respectively.

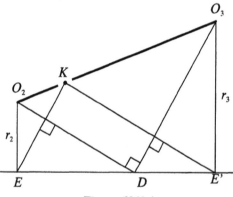

Figure 69/4.4

Notice that if the centres of k_2 and k_3 are O_2 and O_3, then O_2D and O_3D are orthogonal to EF and $E'F'$, respectively (*Figure 1969/4.4*). Since they also bisect the angles at EDF and $E'DF'$, the interval O_2D is orthogonal to O_3D. Moreover, since O_2ED and $DE'O_3$ are similar triangles, their altitudes at E and E' divide the hypotenuse and the interval O_2O_3 in the same manner. This shows that $K = O_1$ is on O_2O_3, which proves our assertion.

Remarks. 1. Notice that our second solution uses only similarities. It is not true in general that O_1 is the midpoint of O_2O_3, as it was deduced in the first solution.

2. Another specialization of the same general problem will be discussed in Problem 1978/4.

1969/5.   *Given $n > 4$ points on the plane (no three collinear), prove that there are at least $\binom{n-3}{2}$ convex quadrilaterals with vertices amongst the given points.*

First solution.   The convex hull of the points ([18]) contains at least three of them as vertices; let us denote such a triple by A, B and C. Add two more points, called X and Y to the above triple. The line XY does not pass through vertices of the triangle ABC and intersects at most two of its edges.

Suppose that it is disjoint from AB. In this case A, B, X, Y form a convex quadrilateral, since otherwise their convex hull is a triangle, implying that XY intersects the interval AB.

Therefore for any two points of the given set minus $\{A, B, C\}$ we can add two of A, B and C such that the resulting quadrilateral is convex. Since different point pairs give rise to different quadrilaterals (because their vertices are different), the number of convex quadrilaterals found in this way is at least

$$\binom{n \perp 3}{2}.$$

Second solution. We start with the fundamental lemma of the subject: from five given points (no three collinear) we can always choose four which form a convex quadrilateral. The proof is similar to the reasoning given above: if the convex hull of the five points is a pentagon or a quadrilateral then four vertices of them will provide the desired convex quadrilateral. In case the convex hull is a triangle, say ABC, then the two inner points X and Y and the edge of ABC disjoint from the line XY — say AB — provides the four points forming a convex quadrilateral (*Figure 1969/5.1*).

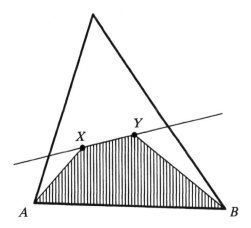

Figure 69/5.1

There are $\binom{n}{5}$ choices for five-tuples of points among n points, according to the above said each tuple determines a convex quadrilateral. This gives $\binom{n}{5}$ convex quadrilaterals; notice, however, that the same quadrilateral might be given by many five-tuples. In fact, there are $n \perp 4$ such tuples determining the same quadrilaterals, since the fifth point can be chosen arbitrarily. In conclusion this argument gives that there are at least

$$\frac{1}{n-4}\binom{n}{5} = \frac{1}{5}\binom{n}{4}$$

different convex quadrilaterals with vertices amongst the given points. This bound is even better than the one required by the problem, since we will show that

$$\frac{1}{5}\binom{n}{4} \geq \binom{n-3}{2}.$$

This inequality is equivalent to $n(n-1)(n-2) \geq 60(n-4)$, which can be verified for $n = 5, 6, 7, 8$ directly; for $n \geq 9$ we have $n(n-1) > 60$ and $n-2 > > n-4$, hence $n(n-1)(n-2) > 60(n-4)$ obviously follows.

Remark. This problem represents one of the most popular circle of problems in convex geometry. A generalization of our starting lemma goes as follows:

For any k there exists $Z(k)$ such that if $n \geq Z(k)$ then for n points in the plane in general position there are k which form a convex k-gon.

The exact value of $Z(k)$, however, is surprisingly hard to determine. We only have estimates for it, for example

$$Z(k) \leq \binom{2k-4}{k-2} + 1.$$

A long-standing conjecture due to P. Erdős and G. Szekeres asserts that $Z(k) = 2^{k-2} + 1$. For $k = 4$ this is exactly the result we found above, and it is known that $Z(5) = 9$.

These kind of problems share many common features with Ramsey-type problems ([13]).

1969/6. *For given real numbers $x_1, x_2, y_1, y_2, z_1, z_2$ satisfying $x_1 > 0$, $x_2 > 0$, $x_1 y_1 - z_1^2 > 0$ and $x_2 y_2 - z_2^2 > 0$, prove that*

(1)
$$\frac{8}{(x_1 + x_2)(y_1 + y_2) - (z_1 + z_2)^2} \leq \frac{1}{x_1 y_1 - z_1^2} + \frac{1}{x_2 y_2 - z_2^2}.$$

Give necessary and sufficient conditions for equality.

First solution. Let us denote the denominators appearing in (1) by A, A_1 and A_2, respectively. The following relation will be crucial for us:

(2) $$A = A_1 + A_2 + x_1 y_2 + x_2 y_1 - 2z_1 z_2.$$

Applying $x + y \geq 2\sqrt{xy}$ $(x > 0, y > 0)$ this implies

$$A = A_1 + A_2 + \frac{x_1}{x_2}(A_2 + z_2^2) + \frac{x_2}{x_1}(A_1 + z_1^2) - 2z_1 z_2,$$

$$A = A_1 + A_2 + \frac{x_1}{x_2}A_2 + \frac{x_2}{x_1}A_1 + \left(z_1\sqrt{\frac{x_2}{x_1}} \perp z_2\sqrt{\frac{x_1}{x_2}}\right)^2.$$

Since the resulting value is positive, it shows that

(3) $\qquad\qquad\qquad\qquad A > 0,$

and (since $x + y \geq 2\sqrt{xy}$) this provides:

(4) $\quad A \geq A_1 + A_2 + \frac{x_1}{x_2}A_2 + \frac{x_2}{x_1}A_1 \geq A_1 + A_2 + 2\sqrt{A_1 A_2} = \left(\sqrt{A_1} + \sqrt{A_2}\right)^2.$

It is not hard to see that equality holds only in case

(5) $\qquad\qquad z_1\sqrt{\frac{x_2}{x_1}} = z_2\sqrt{\frac{x_1}{x_2}}$ and $\frac{x_1}{x_2}A_2 = \frac{x_2}{x_1}A_1.$

Using the above notations (1) can be rewritten as

$$\frac{8}{A} \leq \frac{1}{A_1} + \frac{1}{A_2}, \quad \text{or equivalently } A \geq \frac{8A_1 A_2}{A_1 + A_2}.$$

In order to verify this last inequality (according to (4)) it is enough to show that

$$\left(\sqrt{A_1} + \sqrt{A_2}\right)^2 \geq \frac{8A_1 A_2}{A_1 + A_2}, \quad \text{i.e. } \left(\frac{\sqrt{A_1} + \sqrt{A_2}}{2}\right)^2 \cdot \frac{A_1 + A_2}{2} \geq A_1 A_2.$$

Based on the inequality between the arithmetic and geometric means we have

$$\left(\frac{\sqrt{A_1} + \sqrt{A_2}}{2}\right)^2 \geq \sqrt{A_1 A_2} \text{ and } \frac{A_1 + A_2}{2} \geq \sqrt{A_1 A_2}.$$

Equality holds if and only if $A_1 = A_2$. Together with (5) it gives $x_1^2 = x_2^2$, consequently $x_1 = x_2$ is necessary for equality. Based on (5) these conditions imply $z_1 = z_2$; now $A_1 = A_2$ forces $y_1 = y_2$. But $x_1 = x_2$, $y_1 = y_2$, $z_1 = z_2$ obviously imply equality in (1), hence we determined the necessary and sufficient condition.

Second solution. With the notations of the above solution we note first that the discriminants of the quadratic equations

$$F_1(t) = x_1 t^2 + 2z_1 t + y_1,$$
$$F_2(t) = x_2 t^2 + 2z_2 t + y_2,$$
$$F(t) = F_1(t) + F_2(t) = (x_1 + x_2)t^2 + 2(z_1 + z_2)t + (y_1 + y_2)$$

are $\perp 4A_1$, $\perp 4A_2$ and $\perp 4A$, respectively. Note furthermore that if $a > 0$, then $ax^2 + bx + c$ attains its minimum at $\perp\frac{b}{2a}$ with minimum value $\perp\frac{D}{4a}$ where D is the discriminant of the quadratic equation.

This implies

(5) $$\min F_1(t)=\frac{A_1}{x_1},\quad \min F_2(t)=\frac{A_2}{x_2},\quad \min F(t)=\frac{A}{x_1+x_2}.$$

Since

(6) $$\min F(t)\geq \min F_1(t)+\min F_2(t),$$

(5) gives

$$\frac{A}{x_1+x_2}\geq \frac{A_1}{x_1}+\frac{A_2}{x_2}.$$

Using (3) this inequality is equivalent to

$$\frac{8}{A}\leq \frac{8}{(x_1+x_2)\left(\frac{A_1}{x_1}+\frac{A_2}{x_2}\right)}.$$

We would like to prove that $\dfrac{8}{A}\leq \dfrac{1}{A_1}+\dfrac{1}{A_2}$; according to the above said it is enough to show that

$$\frac{1}{A_1}+\frac{1}{A_2}\geq \frac{8}{(x_1+x_2)\left(\frac{A_1}{x_1}+\frac{A_2}{x_2}\right)}.$$

Since all the quantities appearing in the formula are positive, it is equivalent to

$$\left(\frac{1}{A_1}+\frac{1}{A_2}\right)\left(\left(1+\frac{x_2}{x_1}\right)A_1+\left(1+\frac{x_1}{x_2}\right)A_2\right)\geq 8,$$

$$2+\left(\frac{x_1}{x_2}+\frac{x_2}{x_1}\right)+\left(\frac{A_1}{A_2}+\frac{A_2}{A_1}\right)+\left(\frac{x_2A_1}{x_1A_2}+\frac{x_1A_2}{x_2A_1}\right)\geq 8.$$

This inequality trivially holds since the sum of a positive number and its reciprocal is at least 2; this observation completes the proof of (1).

Equality in this last step holds once $x_1=x_2$ $A_1=A_2$; in (6) we need that the minima of F_1 and F_2 coincide, meaning $\dfrac{z_1}{x_1}=\dfrac{z_2}{x_2}$. Hence $x_1=x_2$ implies $z_1=z_2$ and these equalities together with $A_1=A_2$ yield $y_1=y_2$. In conclusion, equality holds in (1) if and only if $x_1=x_2$, $y_1=y_2$ and $z_1=z_2$.

1970.

1970/1. *M is a point on the side AB of the triangle ABC. Let r_1, r_2 and r denote the radii of the incircles of AMC, BMC and ABC, respectively. ϱ_1, ϱ_2 and ϱ stands for the radii of the excircles of the triangles AMC, BMC and ABC (corresponding to sides AM, BM and AB), respectively. Prove that*
$$\frac{r_1}{\varrho_1}\cdot\frac{r_2}{\varrho_2}=\frac{r}{\varrho}.$$

Solution. The incircle of ABC touches AB at C_0 while the excircle touches the same edge at C_1 (*Figure 1970/1.1*). According to a well-known relation for the tangent intervals (see [19]) we have

(1) $$AC_0 = BC_1 = s \perp a.$$

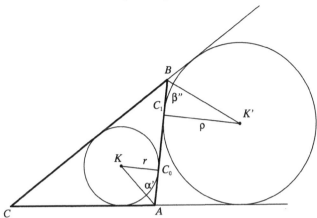

Figure 70/1.1

The centre of the incircle of ABC will be denoted by K, while the centre of the excircle touching AB is K'. If $\angle KAC_0 = \alpha'$ and $\angle K'BC_1 = \beta''$ then

$$r = (s \perp a)\tan \alpha',$$
$$\varrho = (s \perp a)\tan \beta'',$$

and so

(2) $$\frac{r}{\varrho} = \frac{\tan \alpha'}{\tan \beta''}.$$

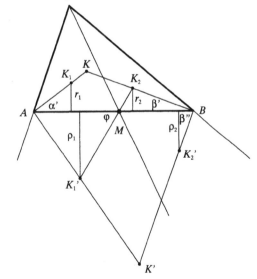

Figure 70/1.2

Furthermore, the centres of the excircle of AMC and the incircle of BMC are denoted by K_1' is K_2, and $\angle K_1'MA = \angle K_2MB = \varphi$. Applying (2) for AMC, and then for the circles of BMC we get (*Figure 19970/1.2*):

$$\frac{r_1}{\varrho_1} = \frac{\tan \alpha'}{\tan \varphi}, \qquad \frac{r_2}{\varrho_2} = \frac{\tan \varphi}{\tan \beta''}.$$

Multiplying the corresponding sides, (2) implies

$$\frac{r_1}{\varrho_1} \cdot \frac{r_2}{\varrho_2} = \frac{\tan \alpha'}{\tan \varphi} \cdot \frac{\tan \varphi}{\tan \beta''} = \frac{\tan \alpha'}{\tan \beta''} = \frac{r}{\varrho},$$

which is exactly what we wanted to prove.

1970/2.   *Real numbers x_i $(i=0,1,\ldots,n)$ with $0 \le x_i < b$ and $x_n > 0$, $x_{n-1} > 0$ are given. If $x_n x_{n-1} \ldots x_1 x_0$ represents the number A_n base a and B_n base b whilst $x_{n-1} \ldots x_1 x_0$ represents A_{n-1} base a and B_{n-1} base b, then prove that $a > b$ holds if and only if*

(1)
$$\frac{A_{n-1}}{A_n} < \frac{B_{n-1}}{B_n}$$

First solution.   According to the notations above

$$A_n = a^n x_n + a^{n-1} x_{n-1} + \ldots + a x_1 + x_0, \quad B_n = b^n x_n + b^{n-1} x_{n-1} + \ldots + b x_1 + x_0,$$

$$A_{n-1} = A_n \perp a^n x_n, \qquad\qquad B_{n-1} = B_n \perp b^n x_n.$$

Easy to see that (1) is equivalent to

(2)
$$A_n B_{n-1} \perp A_{n-1} B_n > 0.$$

Perform the substitutions indicated above and get

$$A_n B_{n-1} \perp A_{n-1} B_n = A_n B_n \perp A_n b^n x_n \perp A_n B_n + B_n a^n x_n =$$

$$= a^n x_n (b^n x_n + b^{n-1} x_{n-1} + \ldots + b x_1 + x_0) \perp$$

$$\perp b^n x_n (a^n x_n + a^{n-1} x_{n-1} + \ldots + a x_1 + x_0) =$$

$$= x_n x_{n-1} a^{n-1} b^{n-1} (a \perp b) + x_n x_{n-2} a^{n-2} b^{n-2} (a^2 \perp b^2) + \ldots$$

$$\ldots + x_n x_1 ab(a^{n-1} \perp b^{n-1}) + x_n x_0 (a^n \perp b^n).$$

Since in this last expression the signs of the terms in the parentheses are the same, (2) is satisfied if and only if $a > b$.

Second solution.   Let us introduce the following notations:

$$f(t) = x_n t^n + x_{n-1} t^{n-1} + \ldots + x_1 t + x_0,$$

$$g(t) = \frac{f(t) \perp x_n t^n}{f(t)} = 1 \perp \frac{x_n t^n}{f(t)}.$$

Since x_0, x_1, ..., x_{n-2} are nonnegative and x_{n-1}, x_n are positive, for positive values of t the functions $f(t)$,

$$\frac{f(t)}{t^n} = x_n + x_{n-1} \frac{1}{t} + x_{n-2} \frac{1}{t^2} + \ldots + x_0 \frac{1}{t^n}$$

and $\dfrac{x_n t^n}{f(t)}$ are strictly increasing while $g(t)$ is strictly decreasing.

Notice that $g(a) = \dfrac{A_{n-1}}{A_n}$ and $g(b) = \dfrac{B_{n-1}}{B_n}$, therefore $a > b$ is equivalent to $g(a) < g(b)$, meaning

$$\frac{A_{n-1}}{A_n} < \frac{B_{n-1}}{B_n}.$$

1970/3.   *The real numbers a_0, a_1, a_2, ..., a_n, ... satisfy*

(1)
$$1 = a_0 \le a_1 \le a_2 \le \ldots \le a_n \le \ldots$$

We define the sequence $b_1, b_2, \ldots, b_n, \ldots$ *as*

(2) $$b_n = \sum_{k=1}^{n}\left(1 - \frac{a_{k-1}}{a_k}\right)\frac{1}{\sqrt{a_k}}.$$

I. *Prove that* $0 \le b_n < 2$ *holds for all* n.

II. *Given* c *satisfying* $0 \le c < 2$, *prove that we can find* $a_0, a_1, \ldots, a_n, \ldots$ *(satisfying (1)) so that infinitely many of the corresponding* b_n *are greater than* c.

First solution. I. Since $\dfrac{a_{k-1}}{a_k} \le 1$ and therefore $1 - \dfrac{a_{k-1}}{a_k} \ge 0$, the terms in the sum defining b_n are nonnegative, and so b_n is nonnegative. One term in the sum under (2) can be rewritten as

$$\left(1 - \frac{a_{k-1}}{a_k}\right)\frac{1}{\sqrt{a_k}} = \left(1 + \sqrt{\frac{a_{k-1}}{a_k}}\right)\left(1 - \sqrt{\frac{a_{k-1}}{a_k}}\right)\frac{1}{\sqrt{a_k}} \le$$

$$\le \frac{2}{\sqrt{a_k}}\left(1 - \sqrt{\frac{a_{k-1}}{a_k}}\right) \le \frac{2}{\sqrt{a_{k-1}}}\left(1 - \sqrt{\frac{a_{k-1}}{a_k}}\right) = 2\left(\frac{1}{\sqrt{a_{k-1}}} - \frac{1}{\sqrt{a_k}}\right).$$

Applying this estimate for (2) we get

$$b_n \le 2\left(\frac{1}{\sqrt{a_0}} - \frac{1}{\sqrt{a_1}} + \frac{1}{\sqrt{a_1}} - \frac{1}{\sqrt{a_2}} + \ldots - \frac{1}{\sqrt{a_n}}\right) = 2\left(1 - \frac{1}{\sqrt{a_n}}\right) < 2,$$

proving Part I. of the problem.

II. Choose $a_0, a_1, \ldots$ to be a geometric sequence with $a_0 = 1$ and quotient $\dfrac{1}{q^2}$ $(0 < q < 1)$. Since $\dfrac{1}{q^2} > 1$, the sequence is increasing, therefore it satisfies (1). Now (2) takes the form

$$\left(1 - \frac{q^{2k}}{q^{2k-2}}\right)q^k = (1 - q^2)q^k,$$

hence

$$b_n = (1 - q^2)\sum_{k=1}^{n} q^k = (1 - q^2)\frac{q(1 - q^n)}{1 - q} = q(1+q)(1 - q^n) > 2q^2(1 - q^n).$$

Since $0 < q < 1$, the sequence q^n converges to 0, and so $1 - q^n$ converges to 1. This implies that for some n_0 the equation $1 - q^n > q$ is satisfied once $n > n_0$. This shows that

(3) $$b_n > 2q^3.$$

Fixing $q = \sqrt[3]{\dfrac{c}{2}}$, $0 \le c < 2$ implies $0 \le q < 1$, hence (3) implies that there is n_0 such that for $n > n_0$

$$b_n > 2q^3 = c,$$

concluding the solution.

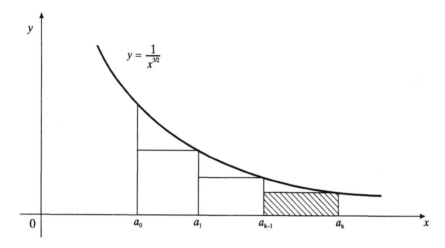

Figure 70/3.1

Remark. It is not hard to guess the origin of this problem; with this guess we get an alternative solution:

A term in (2) can be written as

(4)
$$\frac{a_k \perp a_{k-1}}{a_k^{\frac{3}{2}}}.$$

Let us fix $a_0, a_1, \ldots, a_k, \ldots$ on the x-axis. Now (4) becomes the area of the rectangle with two vertices being equal to a_{k-1}, a_k, and the third vertex is on the curve $y = \dfrac{1}{x^{\frac{3}{2}}}$ (*Figure 1970/3.1*). The value b_n is the sum approximating the area under the curve from below. Therefore b_n is less than the area under the curve from a_0 to a_n, i.e.,

(5)
$$b_n < \int_{a_0}^{a_n} x^{-\frac{3}{2}} dx = \left[\perp \frac{2}{\sqrt{x}}\right]_{a_0}^{a_n} = 2 \perp \frac{2}{\sqrt{a_n}} < 2.$$

This observation verifies Part I. of the problem. The geometric content of Part II. is also clear, but its rigorous proof requires a delicate approximation of the integral. It can be done, but does not provide a simpler result than the one discussed above.

1970/4. *Find all positive integers n such that the set $\{n, n+1, n+2, n+3, n+4, n+5\}$ can be partitioned into two subsets so that the product of the numbers in each subset is equal.*

First solution. We will show that there is no n satisfying the requirements of the problem. Among positive integers every pth is divisible by p and between any two numbers divisible by p there are at least $p \perp 1$ numbers. Therefore there are no two among the given six numbers divisible by 7. If the partition is possible, the only prime factors in the numbers are 2, 3 and 5, since the prime factors in the two sets should be the same.

The prime 5 can appear only in n and $n+5$, hence the prime factorizations of $n+1$ $n+2$, $n+3$ and $n+4$ may contain only 2 and 3 as prime factors. Two among the four numbers are odd, hence divisible only by 3. Their difference is necessarily 2, hence 3 cannot divide both. This contradiction shows that there is no such n.

Second solution. Suppose that there is a partition of the six numbers with the required property. There is no multiple of 7 among our numbers since among six consecutive integers there is at most one divisible by 7, but we need one in each subset of the partition. Therefore $n \pm 1$ is divisible by 7, hence the mod 7 residues are 1, 2, 3, 4, 5, 6. Now the mod 7 residue of the product $6! = 120$ of the six numbers is equal to 6. If the products in the subsets of the partition are equal, their mod 7 residues coincide as well; let this common value be equal to m. The above reasoning shows that mod 7 we have the equality $m^2 = 6$. The mod 7 residue of m can take the values 0, 1, 2, 3, 4, 5, 6, the mod 7 residues of the squares of these are 0, 1, 4, 2, 2, 4, 1, respectively. Since 6 is not among them, the required partitioning is impossible.

Third solution. We will show that if $n \neq 1$ then there is one among the six numbers containing a prime factor in its prime decomposition which is at least 7. Since there is at most one such number among the six chosen one, this argument again shows that the desired partition cannot exist.

The six numbers contain one of the form $6k+1$ and another of the form $6k \pm 1$: there is one which is divisible by six and if it is not n or $n+5$ then the assertion is trivial. Now if n is divisible by 6, then $n+1$ and $n+5$, if $n+5$ is divisible by 6 then n and $n+4$ will admit the promised form. $6k+1$ and $6k \pm 1$ have only odd divisors and 3 is not among them. Only one of the two numbers is divisible by 5 since their difference is 2 or 4. This shows that the other — since it is not equal to 1 — contains a prime factor not less than 7. This final observation concludes the solution.

Remarks. 1. Another way to solve the problem is that we write down all possible partitions of the six numbers, and after computing the products we show that the two resulting numbers are never equal.

2. Even this simple problem has deep number theoretic background. In the second solution we utilized the fact that $6! \equiv 6 \equiv \pm 1 \pmod{7}$. This is a special case of the famous Wilson congruence theorem, which asserts that

$$(p \pm 1)! \equiv \pm 1 \pmod{p}$$

holds if and only if p is a prime.

3. In our third solution we showed that there exists a prime dividing only one of the given six numbers. The following theorem — frequently called the Sylvester–Schur theorem — is closely related to this fact: if $n \geq 2k$ then $\binom{n}{k}$ admits a prime factor greater than k.

Since $\binom{n}{k} = \dfrac{(n - k + 1)(n - k + 2)\ldots(n - 1)n}{k!}$, this means that the ratio is divisible by a prime p greater than k. Obviously the denominator is not divisible by any prime greater than k. Therefore the product (of k terms) in the numerator must contain a term which is divisible by p. Since the numerator is the product of k consecutive numbers, there is at most one divisible by p. In particular this means that if $n > 6$ and so $n + 5 \geq 2 \cdot 6$, then

$$n(n + 1)(n + 2)(n + 3)(n + 4)(n + 5)$$

is divisible by a prime which is greater than 6; this is exactly what we used in our third solution.

The Sylvester–Schur theorem implies that for k consecutive numbers there is a prime dividing exactly one of them. It then implies that k consecutive positive integers can never by partitioned into two subsets with equal products.

The most well-known special case of the Sylvester–Schur theorem is when $n = 2k$: in this case the theorem states that for $n \geq 1$ there is a prime number between n and $2n$ ($n < p < 2n$). This statement is also called the Chebyshev theorem.

1970/5. *In the tetrahedron $ABCD$ the angle $\angle BDC$ is a right angle and the foot of the perpendicular from D to ABC is the intersection of the altitudes of ABC. Prove that*

(1) $$(AB + BC + CA)^2 \leq 6(AD^2 + BD^2 + CD^2).$$

When do we have equality?

Some preparatory remarks. In all solutions we show that the edges at D are perpendicular to each other, hence we have a so-called right angled tetrahedron. After verifying this, the proof of (1) proceeds as follows: let $AB = c$, $BC = a$, $CA = b$, $DA = a_1$, $DB = b_1$, $DC = c_1$. The Pythagorean theorem applied to the faces ABD, BCD, CAD (*Figure 1970/5.1*) yields

(2) $$a^2 = b_1^2 + c_1^2, \qquad b^2 = c_1^2 + a_1^2, \qquad c^2 = a_1^2 + b_1^2.$$

After summing these equalities and multiplying both sides by 3 we get

$$6(a_1^2 + b_1^2 + c_1^2) = 3(a^2 + b^2 + c^2) =$$

$$= (a + b + c)^2 + (a - b)^2 + (b - c)^2 + (c - a)^2 \geq (a + b + c)^2,$$

which proves (3). Equality holds only in case $a = b = c$, based on (2) this implies $a_1 = b_1 = c_1$.

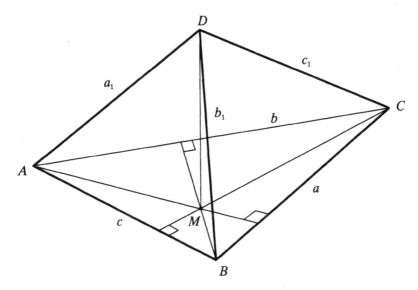

Figure 70/5.1

First solution. Let the intersection of the altitudes in ABC be denoted by M. DM is perpendicular to the plane of ABC and therefore to CA: $CA \perp DM$. We also know that $CA \perp BM$ since BM is an altitude, hence CA is perpendicular to two (nonparallel) lines in the plane of DBM. This implies that it is perpendicular to all lines in the plane at hand, in particular $CA \perp DB$. This shows that DB is perpendicular to two (nonparallel) lines in the plane of CAD (namely to CA and DC), hence it is perpendicular to all lines in this plane. This immediately gives $DB \perp DA$. Changing the roles of B and C we get $DC \perp DA$, hence the edges at the vertex D are pairwise perpendicular, and this concludes the solution.

Second solution. Let $\mathbf{a}$, $\mathbf{b}$, $\mathbf{c}$ and $\mathbf{d}$ denote the vectors pointing from M to the vertices A, B, C and D. We recall that two intervals are perpendicular if and only if the corresponding scalar product vanishes.

DM is perpendicular to the altitudes of ABC:

(4) $$\mathbf{da} = \mathbf{db} = \mathbf{dc} = 0.$$

MB is perpendicular to AC and MC is perpendicular to AB:

(5) $$\mathbf{b}(\mathbf{c} \perp \mathbf{a}) = 0, \quad \text{i.e.} \quad \mathbf{bc} = \mathbf{ab},$$

and

(6) $$\mathbf{c}(\mathbf{a} \perp \mathbf{b}) = 0, \quad \text{i.e.} \quad \mathbf{ac} = \mathbf{bc}.$$

DB is perpendicular to DC, hence

$$(\mathbf{b} \perp \mathbf{d})(\mathbf{c} \perp \mathbf{d}) = 0, \quad \text{i.e.} \quad \mathbf{bc} + \mathbf{d}^2 \perp \mathbf{dc} \perp \mathbf{db} = 0,$$

and so according to (4) we get

(7) $$\mathbf{bc} + \mathbf{d}^2 = 0.$$

We want to show that DB and DC are perpendicular to DA. In the spirit of (7) these statements are equivalent to

$$\mathbf{ab} + \mathbf{d}^2 = 0, \quad \text{and} \quad \mathbf{ac} + \mathbf{d}^2 = 0.$$

These latter equations, however, follow directly from (7), since (5) and (6) show

$$\mathbf{bc} = \mathbf{ab} = \mathbf{ac}.$$

Remarks. A tetrahedron with the above property is called an orthocentric tetrahedron; the altitudes of these solids pass through a common point (see [20]). In our case the tetrahedron has an additional special property: it admits a vertex at which the edges are pairwise perpendicular to each other. This property explains the name: right angled tetrahedron. In a certain sense such a solid can be interpreted as the three dimensional generalization of a right angled triangle. For example, there is a relation reminiscent to the usual Pythagorean theorem: the squares of areas of the three faces with right angles is equal to the square of the area of the acute face.

1970/6. *Given 100 coplanar points (no three collinear), consider all triangles with vertices among the given points and prove that at most 70 % of these triangles have all angles acute.*

First solution. First we verify the statement for 5 points (we always assume that no three are collinear). Consider the convex hull of the given points; it is a triangle, a quadrilateral or a pentagon.

a) If the convex hull is the triangle ABC (*Figure 1970/6.1a*) then it contains the remaining two points, D and E, in its interior. The intervals joining D with the vertices of the triangle form 3 angles, and their sum is $360°$. Therefore at most two of these angles are not acute. This shows the existence of two obtuse triangles. The same argument applies to E, hence there are *at least* 4 obtuse triangles on the given 5 points.

a)

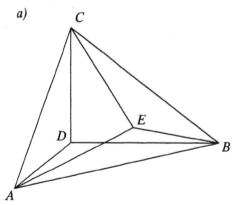

b)

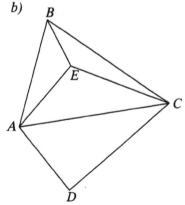

Figure 70/6.1a   Figure 70/6.1b

b) If the convex hull is a quadrilateral, denoted by $ABCD$ (*Figure 1970/6.1b*), then the fifth point E is in the interior of, say, the triangle ABC. At least one angle of the quadrilateral is not acute, hence the edges of this angle determine an obtuse triangle. As before, the intervals EA, EB, EC determine at least two obtuse angles, giving rise to further two obtuse triangles. In summary, we found *at least* three obtuse triangles.

c) Finally, assume that the convex hull is equal to the pentagon $ABCDE$. Since the sum of the angles in the pentagon is 540°, it has at least two obtuse angles (*Figure 1970/6.1c*). If these obtuse angles are at neighbouring vertices A and B, then AEB and CBA are obtuse triangles. The quadrilateral $EDCB$ also has an obtuse angle, and the corresponding triangle is obtuse again. This shows the existence of *at least* three obtuse triangles in this case again.

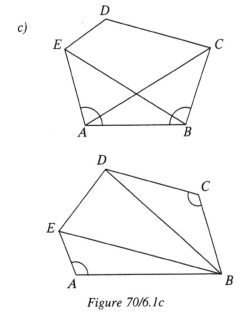

Figure 70/6.1c

If the two obtuse angles are not neighbouring, say these are at the vertices A and C, then besides the triangles at these vertices the quadrilateral $ACDE$ provides an obtuse triangle (since at least one of its angles is obtuse). Once again we found *at least* three obtuse triangles.

In conclusion we deduced that among the $\binom{5}{3} = 10$ triangles determined by the 5 given points there are at least 3 obtuse, in conclusion the number of acute triangles is at most 7; i.e., at most 70 % of all the triangles.

Now let the number of points be equal to $n > 5$. We can choose five tuples from these n points in $\binom{n}{5}$ ways, and each such five tuple contains at least three obtuse triangles. At first glance this seems to give at least $3\binom{n}{5}$ obtuse triangles.

Notice, however, that each such triangle is counted with multiplicity, we counted them in as many five tuples as many extensions of the three points (the vertices) to a five tuple exist. There are $\binom{n-3}{2}$ such extensions, hence the number of obtuse triangles we have found in this way is at least

$$\frac{3\binom{n}{5}}{\binom{n-3}{2}}.$$

This shows that the number of acute triangles is at most $\binom{n}{3} - \frac{3\binom{n}{5}}{\binom{n-3}{2}}$. The ratio of this number and the total number of triangles is

$$\frac{\binom{n}{3} - \frac{3\binom{n}{5}}{\binom{n-3}{2}}}{\binom{n}{3}} = 1 - \frac{3\binom{n}{5}}{\binom{n}{3}\binom{n-3}{2}} = 1 - \frac{3}{10} = 0{,}7 = 70\ \%.$$

Thus we proved the assertion for every $n \geq 5$.

Second solution. The idea is the following: we will show that the ratio of the number of acute triangles and the number of all triangles on n points decreases once n increases. Since the ratio does not exceed 70 % for $n=5$, this monotonicity shows that the same holds for all $n > 5$.

Let $M(n)$ denote the maximum number of acute triangles determined by $n \geq 3$ coplanar points. Let furthermore A_1, A_2, ..., A_{n+1} denote $n+1$ points on the plane (no three are collinear). Delete A_i; the number of acute triangles on the remaining n points is denoted by k_i. The sum $k_1 + k_2 + \ldots + k_{n+1}$ will give the acute triangles on A_1, A_2, ..., A_{n+1} in a way that, for example, the triangle $A_j A_k A_l$ is counted in this sum as many times as many ways we can add $n-3$ points to A_j, A_k, A_l from the remaining $n-2$. This means that each such triangle is counted with multiplicity $n-2$. The number of acute triangles determined by A_1, A_2, ..., A_{n+1} will be denoted by h. Obviously $h \leq M(n+1)$ since h corresponds to a specific set while $M(n+1)$ refers to all sets of $n+1$ elements. Therefore

$$\frac{k_1 + k_2 + \ldots + k_{n+1}}{n-2} = h.$$

By definition we have $k_i \leq M(n)$, hence

$$k_1 + k_2 + \ldots + k_{n+1} = (n-2)h \leq (n+1)M(n),$$

i.e.

$$h \leq \frac{n+1}{n-2}M(n).$$

This inequality holds for any set A_1, A_2, ..., A_{n+1} on the plane, in particular for the one in which the number of acute triangles is exactly $M(n+1)$. This shows

$$M(n+1) \le \frac{n+1}{n \perp 2} M(n).$$

Dividing the inequality by $\binom{n+1}{3}$ and taking $\frac{n \perp 2}{n+1} \cdot \binom{n+1}{3} = \binom{n}{3}$ into account we get

(1)
$$\frac{M(n+1)}{\binom{n+1}{3}} \le \frac{M(n)}{\binom{n}{3}}.$$

This last inequality means exactly that the ratio of the acute triangles and the total number of triangles on n points gives a monotone decreasing function when n increases.

If $n = 5$, then — as was shown by the first solution — $M(5) = 7$, and so $\frac{M(5)}{10} = \frac{7}{10} = 0{,}7$. Now (1) implies that for $n \ge 5$

$$\frac{M(n)}{\binom{n}{3}} \le 0{,}7 = 70 \ \%.$$

1971.

1971/1. *Show that the following statement is true for $n = 3$ and 5 and false for all other $n > 2$:*

"*For any real numbers a_1, a_2, ..., a_n the inequality*

$$(a_1 \perp a_2)(a_1 \perp a_3) \ldots (a_1 \perp a_n) + (a_2 \perp a_1)(a_2 \perp a_3) \ldots (a_2 \perp a_n) + \ldots$$
$$\ldots + (a_n \perp a_1)(a_n \perp a_2) \ldots (a_n \perp a_{n-1}) \ge 0.$$

holds".

Solution. For $n = 3$ we have

$$(a_1 \perp a_2)(a_1 \perp a_3) + (a_2 \perp a_1)(a_2 \perp a_3) + (a_3 \perp a_1)(a_3 \perp a_2) =$$
$$= a_1^2 + a_2^2 + a_3^2 \perp a_1 a_2 \perp a_2 a_3 \perp a_3 a_1 =$$
$$= \frac{1}{2} \left((a_1 \perp a_2)^2 + (a_2 \perp a_3)^2 + (a_3 \perp a_1)^2 \right) \ge 0,$$

hence the inequality holds.

For $n = 5$ the expression is symmetric in the a_i's hence we can assume that $a_1 \ge a_2 \ge a_3 \ge a_4 \ge a_5$. Rewrite the expression as

$$(a_1 \perp a_2)[(a_1 \perp a_3)(a_1 \perp a_4)(a_1 \perp a_5) \perp (a_2 \perp a_3)(a_2 \perp a_4)(a_2 \perp a_5)] +$$
$$+ (a_1 \perp a_3)(a_2 \perp a_3)(a_4 \perp a_3)(a_5 \perp a_3) +$$
$$+ (a_4 \perp a_5)[(a_1 \perp a_5)(a_2 \perp a_5)(a_3 \perp a_5) \perp (a_1 \perp a_4)(a_2 \perp a_4)(a_3 \perp a_4)].$$

In the first and third terms $a_1 \perp a_2 \geq 0$ and $a_4 \perp a_5 \geq 0$. In the square brackets we have differences of products involving three factors and we have $a_1 \perp a_3 \geq$ $\geq a_2 \perp a_3$, $a_1 \perp a_4 \geq a_2 \perp a_4$, $a_1 \perp a_5 \geq a_2 \perp a_5$, and similarly: $a_1 \perp a_5 \geq a_1 \perp$ $\perp a_4$, $a_2 \perp a_5 \geq a_2 \perp a_4$, $a_3 \perp a_5 \geq a_3 \perp a_4$. Therefore these sums give nonnegative numbers, hence the first and third terms in the above sum are nonnegative. The middle term is simply a product of two nonpositive and two nonnegative numbers, consequently it is nonnegative. Therefore the expression is nonnegative once $n = 5$, hence the statement is verified.

For $n = 4$ consider $a_1 = 0$, $a_2 = a_3 = a_4 = 1$; the value of the expression then becomes $\perp 1$, hence the inequality in the statement is false.

Now for $n > 5$ consider the substitution $a_1 = 1$, $a_2 = a_3 = a_4 = 2$, $a_5 = a_6 =$ $= \ldots = a_n = 0$; the resulting value is $\perp 1$ again. This last example concludes the solution of the problem.

Remark. The following rewording of the problem might shed light on its origin: Let $p'(x)$ be the derivative of the polynomial

$$p(x) = (x \perp a_1)(x \perp a_2) \ldots (x \perp a_k) \qquad (n \geq 2).$$

The inequality

$$\sum_{i=1}^{n} p'(a_i) \geq 0 \qquad (i = 1, 2, \ldots, n)$$

holds for any real n-tuple $a_1, a_2, \ldots, a_n$ exactly in case $n = 3$ and $n = 5$. Hence the problem admits a solution based on the relation between polynomials and their derivatives.

1971/2. *Let P_1 be a convex polyhedron with vertices $A_1, A_2, \ldots, A_9$. Let P_i be the polyhedron obtained from P_1 by a translation that moves A_1 into A_i $(i = 2, 3, \ldots, 9)$. Show that at least two of the polyhedra $P_1, P_2, \ldots, P_9$ have an interior common point.*

Solution. Let Q be a point of P_1 and apply the translation $\overrightarrow{A_1 A_i}$ to it *(Figure 1971/2.1)*, and denote the resulting point by Q'. Fixing the origin at A_1, in vector notation we have $\overrightarrow{A_1 Q'} = \overrightarrow{A_1 Q} + \overrightarrow{A_1 A_i}$. According to the parallelogram rule Q' can also be obtained by reflecting A_1 to the midpoint C of the interval $A_i Q$ (this holds even in case Q in on $A_1 A_i$). Since both Q and A_i are in P_1, and P_1 is convex, it follows that C is in P_1 as well. This shows that Q' can be got by applying an enlargement with centre A_1 and quotient 2 to C.

In summary, we showed that by translation we got points which are in the solid P_1 we have got from P_1 by the above enlargement.

Let the volume of P_1 be denoted by V; therefore P_1' has volume $8V$. Since all the polyhedrons $P_1, P_2, \ldots, P_9$ with volume V are in P_1' of volume $8V$, they

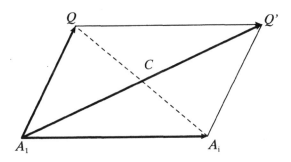

Figure 71/2.1

cannot be disjoint, hence there are at least two different sharing common interior points.

Remark. For a polyhedron with eight vertices the statement is usually false — consider the cube, for example. For all polyhedra with more than 8 vertices, though, the statement turns out to be true.

1971/3. *Prove that we can find an infinite set of positive integers of the form $\{2^n \perp 3\}$ (where n is a positive integer) every pair of which are relatively prime.*

Solution. For $n = 3, 4, 5$ the values of $2^n \perp 3$ are 5, 13 and 29, respectively, and these are relatively prime. We will give a procedure which gives further elements of the desired form and the resulting numbers are relative primes to the previously constructed ones.

Let a_1, a_2, ..., a_k be given with the required properties. Consider $s = a_1 a_2 \ldots a_k$ and take the $s + 1$ numbers

$$2^0, \; 2^1, \; 2^2, \; \ldots, \; 2^s.$$

There are two among them with the same residue mod s, say 2^r and 2^q $(r > q)$. Since $2^r \perp 2^q = 2^q(2^{r-q} \perp 1)$ is divisible by s and s is a product of odd numbers (and so itself is odd), s divides $2^{r-q} \perp 1$, i.e., for some integer e we have $2^{r-q} \perp$ $\perp 1 = es$. Define now

$$a_{k+1} = 4es + 1 = 4 \cdot 2^{r-s} \perp 4 + 1 = 2^{r-s+2} \perp 3.$$

a_{k+1} is obviously relatively prime to a_i $(i \leq k)$, since a common divisor (greater than 1) of a_i and a_{k+1} divides both s and $4es + 1$, which is a contradiction. Finally a_{k+1} is greater than any a_i $(i \leq k)$, since it is greater than their product. Now applying this method an infinite sequence of integers with the desired properties can be constructed.

Remarks. 1. Our argument can be slightly modified as follows: Let

$$a_{k+1} = 2^{\varphi(s)+2} \perp 3$$

where φ is the Euler function $\varphi(n)$ ([21]). According to the Euler congruence theorem $2^{\varphi(s)} \equiv 1 \pmod{s}$ and $2^{\varphi(s)} \perp 1 = es$, where e is an integer; consequently

$$a_{k+1} = 4 \cdot 2^{\varphi(s)} \perp 3 = 4es + 1.$$

This implies $a_{k+1} > a_i$ $(i \leq k)$, furthermore a_{k+1} and a_i are relatively prime since their common divisor (different from 1) divides $4es + 1$, which is a contradiction.

2. The same argument applies to numbers of the form $\{p^n \perp (p+1)\}$ for some prime p.

3. Our construction followed the line first used by Euclid in proving the existence of infinitely many primes: for finitely many prime numbers $p_1, p_2, \ldots, p_k$, consider $p_{k+1} = p_1 p_2 \ldots p_k + 1$. It has to be prime since it is not divisible by any of the p_i's.

1971/4. *All faces of the tetrahedron $ABCD$ are acute triangles. Let X, Y, Z and T be points in the interiors of the segments AB, BC, CD and DA, respectively; and consider the closed path $XYZTX$.*

a) *If*

(1) $$\angle DAB + \angle BCD \neq \angle ABC + \angle CDA,$$

then prove that none of the closed paths $XYZTX$ has minimal length.

b) *If*

(2) $$\angle DAB + \angle BCD = \angle ABC + \angle CDA,$$

then prove that there are infinitely many shortest paths $XYZTX$, each with length

$$2AC \sin \frac{\alpha}{2},$$

where $\alpha = \angle BAC + \angle CAD + \angle DAB$.

Solution. Cut the tetrahedron along the edges AC, AB, BD and unfold it into the plane; this construction yields the lattice of the tetrahedron (*Figure 1971/4.1*). This operation does not change intervals on individual faces, hence the closed path on the tetrahedron has minimal length if and only if the corresponding path on the plane has minimal length.

The union of two faces (since they are acute triangles) form a convex quadrilateral. If, for example, on the quadrilateral $ABDC$ the intervals XY and YZ are not collinear then Y can be chosen to achieve that X, Y and Z are collinear and Y is an interior point of BC. Consequently, in the plane the points of the shortest path $XYZTX$ are collinear; the same is true if we cut our tetrahedron along another edge, say AD.

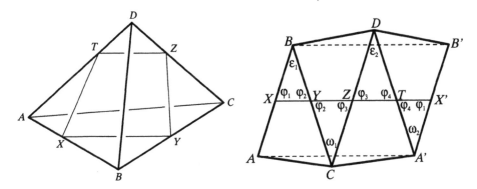

Figure 71/4.1

Assuming that X, Y, Z and T are collinear, we introduce the following notations:

$$\angle AXT = \angle BXY = \varphi_1, \quad \angle BYX = \angle CYZ = \varphi_2,$$
$$\angle CZY = \angle DZT = \varphi_3, \quad \angle DTZ = \angle ATX = \varphi_4.$$

With these notations at hand we have

$$\angle DAB + \angle BCD = 2\pi \perp (\varphi_1 + \varphi_2 + \varphi_3 + \varphi_4),$$
$$\angle ABC + \angle CDA = 2\pi \perp (\varphi_1 + \varphi_2 + \varphi_3 + \varphi_4),$$

which contradicts (1), hence there is no shortest closed path $XYZTX$.

Suppose now that (2) is satisfied. Let the angles at the vertices A, B, C and D be denoted by α, β, γ and δ, respectively. Applying (2) we get

$$\alpha + \gamma = \angle DAB + \angle DAC + \angle CAB + \angle BCD + \angle BCA + \angle ACD =$$
$$= (\angle DAB + \angle BCD) + (\angle CAB + \angle BCA) + (\angle DAC + \angle ACD) =$$
$$= \angle ABC + \angle CDA + \pi \perp \angle ABC + \pi \perp \angle CDA = 2\pi,$$

therefore

(3) $$\alpha + \gamma = 2\pi.$$

The same reasoning shows that (2) implies

(4) $$\beta + \delta = 2\pi.$$

Assume that $\gamma \leq \pi$ and $\delta \leq \pi$. (In the contrary case: (3) and (4) implies that at least one of the pairs (α, δ), (α, β), (γ, β), (γ, δ) satisfies the inequality, if this happens to be (α, β), for example, then cut the tetrahedron and unfold it into the plane in a way that AB plays the role of "central" edge [played by CD on our figure].) This shows that the hexagon $ABDB'A'C$ is convex.

Now (2) implies that AB and $A'B'$ are parallel (see also our remark), and since these intervals are of equal length, we conclude that $AA'B'B$ is a parallelogram. Hence any closed path mapped into an interval XX' parallel to AA' is of minimal length, and all these lengths are equal.

Since $AC = A'C$, the triangle ACA' is isosceles with angle $\angle ACA'$ equal to $\dot{\gamma}$, therefore the length of XX' is

$$XX' = AA' = 2AC \sin \frac{\gamma}{2} = 2AC \sin\left(\pi \perp \frac{\alpha}{2}\right) = 2AC \sin \frac{\alpha}{2},$$

and this is what we wanted to prove.

Remark. The fact that AB and $A'B'$ are parallel can be verified as follows: according to (2), if $\angle DAB = \omega_2$, $\angle BCD = \omega_1$, $\angle ABC = \varepsilon_1$, $\angle ADC = \varepsilon_2$ then

$$\omega_1 + \omega_2 = \varepsilon_1 + \varepsilon_2,$$

hence

$$\perp\varepsilon_1 + \omega_1 \perp \varepsilon_2 + \omega_2 = 0.$$

This means that by rotating a vector parallel to AB with angles $\perp\varepsilon_1$, $+\omega_1$, $\perp\varepsilon_2$ and $+\omega_2$ respectively, its direction remains unchanged, but applying the same transformations to the line AB with centres B, C, D, A' one after another, then it is mapped to $A'B'$. Since its position has not been changed, we get that AB and $A'B'$ are parallel.

1971/5. *Prove that for every positive integer m we can find a finite set of points S in the plane such that for any point A of S, there are exactly m points in S at unit distance from A.*

Solution. For $m = 1$ we can choose S to be the two endpoints of the unit interval; for $m = 2$ the three vertices of an equilateral triangle of unit side length will do. Next we will describe a simple method for producing a set S_{m+1} corresponding to the value $m + 1$ from a set S_m satisfying the requirements of the problem for m.

Translate the points of S_m by a given appropriate unit vector; the resulting points define S'_m. The union of S_m and S'_m now provides a set which has the property we require from S_{m+1}: for each point of S_m and S'_m there is an additional one which is of unit distance.

Not all unit vectors will provide an appropriate S_{m+1}, though. Applying translation by **e** for the equilateral triangle shown by *Figure 1971/5.1* gives a set of type S_3 while doing the same with the vector **e'** is not suitable for our purposes, since there are four points of unit distance from A'.

Let us examine how can we avoid the "bad" translations. A translation is bad if it maps one point of S_m into another point of the same set, or into one for which there are more than one points of S_{m+1} of unit distance.

Construct a circle k_i of radius 1 for each $A_i \in S_m$ as centre. The points of intersection of k_i with the other unit circles are denoted by $P_1, P_2, \ldots, P_k$; these are the points A_i is not supposed to be translated. Therefore we omit $\overrightarrow{A_iP_1}$, $\overrightarrow{A_iP_2}, \ldots, \overrightarrow{A_iP_k}$ from the unit vectors we would like to use in our translation

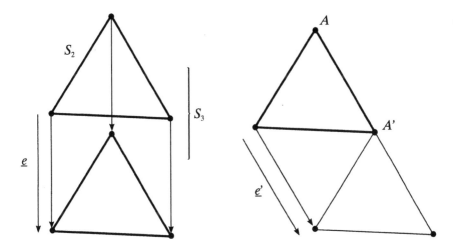

Figure 71/5.1

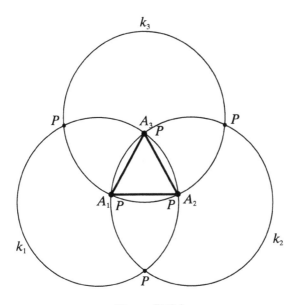

Figure 71/5.2

process. After doing so for every i we exclude finitely many vectors; now all the remaining ones will provide an appropriate set S_{m+1}. On *Figure 1971/5.2* we denoted by P those points corresponding to the vertices A_1, A_2, A_3 yielding unit vectors not suitable for our present purposes.

In conclusion, we showed that starting from any set S_1, by iterating the above process a set S_m for any m with the required properties can be constructed.

1971/6. *Let* $A =$

$$
\begin{matrix}
a_{11} & a_{12} & \cdots & a_{1n} \\
a_{21} & a_{22} & \cdots & a_{2n} \\
\vdots & & & \\
a_{n1} & a_{n2} & \cdots & a_{nn}
\end{matrix}
$$

be a square matrix with all a_{ij} *nonnegative integers. For each* i, j *with* $a_{ij} = 0$ *we have*

(1) $\qquad a_{i1} + a_{i2} + \ldots + a_{in} + a_{1j} + a_{2j} + \ldots + a_{nj} \geq n.$

Prove that the sum of all the elements in the matrix is at least

$$
\frac{n^2}{2}.
$$

First solution. (1) simply means that if there is a position in the matrix where the corresponding value is zero, then the sum of the elements in that row and column is at least n. Notice that this assumption remains unchanged after changing two rows or two columns or rotating the entire matrix by 90°. (This last operation turns rows into columns and vice versa.)

Add the numbers in each row and column and denote the minimum of the resulting $2n$ numbers by s. We can assume that this minimum is attained by the first row. If $s \geq n$ then the sum of the elements in the matrix is at least n^2, hence the statement is true.

If $s < n$, then the first row contains at least $n \perp s$ zeros. Rearrange the matrix by column changes such that the zeros are at the beginning of the row. In that columns the sum of the elements — according to (1) — is at least $n \perp s$.

Henceforth in the first $n \perp s$ columns (in the ones surely starting with 0) the sum of terms is at least $(n \perp s)(n \perp s) = (n \perp s)^2$. In the remaining s columns the sums are at least s columnwise, giving at least s^2 to the total sum. Hence the sum of all the elements can be estimated by

$$
(n \perp s)^2 + s^2 = \frac{n^2}{2} + \frac{(n \perp 2s)^2}{2} \geq \frac{n^2}{2},
$$

which now solves the problem.

Second solution. Rearrange the matrix using row-row and column-column changes such that the most possible zeros stand in the upper left end of the main diagonal. Let us assume that the maximum number of the zeros in the main diagonal is k. Now partition the matrix into four parts as it is shows by *Figure 1971/6.1*. The upper $k \times k$ minor A has all zeros in the main diagonal, while in D there are no zeros — otherwise a row-row or column-column change would transport an additional zero into the main diagonal.

B and C cannot contain zeros which are symmetric to the main diagonal, otherwise by an appropriate change of two rows a zero can be transported to D.

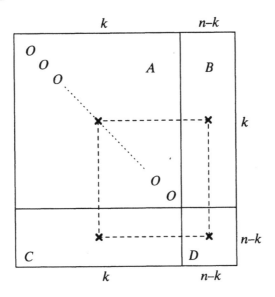

Figure 71/6.1

The sums of the terms in the individual parts will be denoted by $s(A)$, $s(B)$, $s(C)$ and $s(D)$.

Since D contains no zeros, we get $s(D) \geq (n-k)^2$. Adding up the rows and columns corresponding to diagonal positions in A (because of (1)) we get at least kn. In this sum the elements of A appear twice while the elements of B and C appear once. Therefore

$$2s(A) + s(B) + s(C) \geq nk.$$

Now add the elements of B and C in the following way: add an element in B first to its reflection to the main diagonal (which is in C). Since there are no opposite zeros, this procedure provides at least 1 in every position in C, implying

$$s(B) + s(C) \geq k(n-k).$$

Adding $2s(D) \geq 2(n-k)^2$ to these two inequalities we get

$$2\left(s(A) + s(B) + s(C) + s(D)\right) \geq nk + k(n-k) + 2(n-k)^2 = n^2 + (n-k)^2.$$

This shows

$$s(A) + s(B) + s(C) + s(D) \geq \frac{n^2}{2},$$

which coincides with the inequality we wanted to demonstrate.

Remark. The estimate cannot be improved as the following example shows: Let $n = 2k$, and put $\dfrac{n}{2}$ in every position of the main diagonal and zero everywhere else.

1972.

1972/1. *Given any set of ten distinct numbers in the range* 10, 11, ..., 99, *prove that we can always find two disjoint subsets with the same sum.*

Solution. A set of ten elements has $2^{10} \perp 2 = 1022$ proper subsets. The sum of at most 9 numbers, each of two digits is at least 10 and at most

$$91 + 92 + \ldots + 99 = 855;$$

hence there are at most 846 possible values for such a sum. Therefore by forming sums containing one, two, three, ... , nine terms from the given ten numbers, there are two which are equal. If there are common terms in these two sums, then we can delete them and we still have equality. It cannot happen that during this procedure one sum becomes zero, since then the same should happen to the other one, although the two sums are not composed from the same terms. Hence there are always two disjoint subsets with equal sum.

1972/2. *Given $n > 4$, prove that every cyclic quadrilateral can be dissected into n cyclic quadrilaterals.*

Remarks before the solution. The cyclic property of a quadrilateral depends only on its angles, more precisely if it admits two opposite angles with sum 180° then the quadrilateral is cyclic. Therefore if we perform changes on a cyclic quadrilateral leaving the angles unchanged, the result will be cyclic as well.

Every symmetric trapezium is a cyclic quadrilateral and we can dissect it into an arbitrary number of cyclic quadrilaterals by lines parallel to its base. Therefore the statement is obviously true for a symmetric trapezium and hence for rectangles — so we can disregard them in our subsequent arguments.

First solution. Let the angle of the cyclic quadrilateral $ABCD$ at A be acute (since $\angle A + \angle C = 180°$, and not all angles of the quadrilateral are right angles, such vertex exists). Reflect $ABCD$ to the bisector of $\angle A$ and then shrink it until it gets inside $ABCD$. The resulting quadrilateral will be denoted by $AB'C'D'$. According to our remark above it is cyclic (*Figure 1972/2.1*). The parallels to AD and AB passing through C intersect CD and BC in X and Y, respectively. $B'C'XD$ is a symmetric trapezium since the angles at B' and D coincide. The same reasoning shows that $D'C'YB$ is a symmetric trapezium. Finally, $XC'YC$ is a cyclic quadrilateral since its angles coincide with the angles of $ABCD$.

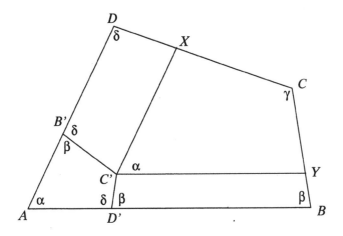

Figure 72/2.1

In this way $ABCD$ has been dissected into four cyclic quadrilaterals; now dissecting one of the symmetric trapeziums further, the number of pieces can be increased arbitrarily with keeping the condition that all quadrilaterals are cyclic.

Second solution. The two obtuse angles of the cyclic quadrilateral cannot be opposite since the sum of opposite angles is equal to $180°$. Let $\angle C = \gamma$ and $\angle D = \delta$ two neighbouring obtuse angles. The section $A'B'$ parallel to DC dissects the original quadrilateral into a cyclic quadrilateral $A'B'BA$ and a trapezium $A'B'CD$. If $\gamma = \delta$ then $ABCD$ is a symmetric trapezium, therefore we can assume (by symmetry) that $\gamma > \delta$ (*Figure 1972/2.2*).

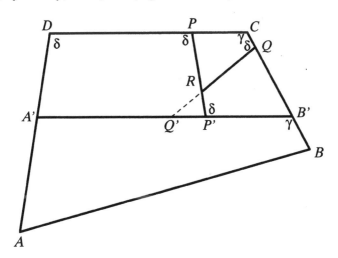

Figure 72/2.2

Choose P and P' on CD and $A'B'$, respectively, in such a way that $\angle DPP' = \delta$, and then Q and Q' on CB' and $A'B'$ such that $\angle CQQ' = \delta$. Now the intervals PP' and QQ' can be translated to intersect each other in an inner point R of the trapezium. In this case the trapezium $A'B'CD$ is dissected into three cyclic quadrilaterals: $A'P'PD$ is a symmetric trapezium since DP and $A'P'$ are parallel and $\angle A'DP = \angle DPP' = \delta$; $QCPR$ is a cyclic quadrilateral since $\angle CQR = \delta$ is equal to the complement of the opposite angle and finally $B'P'RQ$ is a cyclic quadrilateral because $\angle B'P'R = \delta$ is equal to the complement of the opposite angle.

In this way $ABCD$ has been dissected into four cyclic quadrilaterals and the symmetric trapezium can be further dissected into as many cyclic quadrilaterals as one wishes.

Remark. There are numerous different solutions of the problem; the above two share the advantage that they apply to all cases at once. Most of the other solutions distinguish cases according to where the centre of the circumcircle of the cyclic quadrilateral lies.

1972/3. *Let m and n be nonnegative integers. Prove that*

$$\frac{(2m)!(2n)!}{m!n!(m+n)!}$$

is an integer. (According to our conventions $0! = 1$).

First solution.   Introduce the notation

(1) $$f(m,n) = \frac{(2m)!(2n)!}{m!n!(m+n)!}.$$

$f(m,0)$ is an integer for all nonnegative integer m since

$$f(m,0) = \frac{(2m)!}{(m!)^2} = \binom{2m}{m},$$

and the combinatorial interpretation of this expression shows that it is an integer.

The function $f(m,n)$ satisfies the following equation:

(2) $$f(m+1,n) + f(m,n+1) = 4f(m,n).$$

We prove this by substituting the definition of the left hand side:

$$f(m+1,n) + f(m,n+1) = \frac{(2m+2)!(2n)!}{(m+1)!n!(m+n+1)!} + \frac{(2m)!(2n+2)!}{m!(n+1)!(m+n+1)!} =$$

$$= \frac{(2m)!(2n)!}{m!n!(m+n)!} \cdot \frac{4m+2+4n+2}{m+n+1} = 4f(m,n).$$

This verifies (2). After substituting $n := n \perp 1$ and reordering the equation we have

(3) $$f(m,n) = 4f(m,n \perp 1) \perp f(m+1,n \perp 1).$$

Now induction on n implies that $f(m,n)$ is an integer for every m and n: We already saw that it holds for $n = 0$ and arbitrary m. Suppose that $f(m,n \perp 1)$ is an integer for all m, now (3) implies that $f(m, n)$ is an integer; this concludes our first solution.

Second solution. The main idea of this solution is the following: we show that if a prime p divides the denominator then it also divides the numerator, and in addition the exponent of it in the numerator is at least as much as in the denominator. This exactly means that the ratio is an integer.

First we prove a statement originally due to Legendre: the exponent of a prime p in the prime factorization of $n!$ is:

$$\left[\frac{n}{p}\right] + \left[\frac{n}{p^2}\right] + \left[\frac{n}{p^3}\right] + \ldots + \left[\frac{n}{p^k}\right] \qquad (p^k \le n < p^{k+1}).$$

This is true, since among the first n positive integers every pth is divisible by p, hence there are $\left[\frac{n}{p}\right]$ such numbers. Every p^2th is divisible by p^2, and there are $\left[\frac{n}{p^2}\right]$ of those, and so on: there are $\left[\frac{n}{p^k}\right]$ numbers divisible by p^k. Therefore p appears in $n!$ with exponent $\left[\frac{n}{p}\right] + \left[\frac{n}{p^2}\right] + \ldots \left[\frac{n}{p^k}\right]$. Notice that for $p^{k+1} > n$ we have $\left[\frac{n}{p^{k+1}}\right] = 0$, hence there is no mistake by writing these terms in an infinite sum.

We will need one more statement: For a, b nonnegative reals

(4) $$[2a] + [2b] \ge [a] + [b] + [a+b].$$

To prove this let $a = [a] + \alpha$ and $b = [b] + \beta$ where $0 \le \alpha < 1, 0 \le \beta < 1$. If $\alpha + \beta < 1$ then $[a+b] = [a] + [b]$ and so

$$[2a] + [2b] = 2[a] + 2[b] + [2(\alpha + \beta)] \ge 2[a] + 2[b] = [a] + [b] + [a+b].$$

In case $\alpha + \beta \ge 1$ we have either $2\alpha \ge 1$ or $2\beta \ge 1$. Suppose that $2\alpha \ge 1$. Then

$$[a+b] = [a] + [b] + [\alpha + \beta] = [a] + [b] + 1,$$
$$[2a] = 2[a] + [2\alpha] = 2[a] + 1.$$

This implies

$$[2a] + [2b] \ge 2[a] + 1 + 2[b] = [a] + [b] + [a+b],$$

proving the statement. A similar argument concludes the verification of (4) in the case $2\beta \ge 1$.

Now the exponent of p in the numerator of the ratio under examination is

$$A = \left[\frac{2m}{p}\right] + \left[\frac{2m}{p^2}\right] + \ldots + \left[\frac{2m}{p^k}\right] + \left[\frac{2n}{p}\right] + \left[\frac{2n}{p^2}\right] + \ldots + \left[\frac{2n}{p^k}\right].$$

Choose k such that p^{k+1} is greater than $2m$ and $2n$.

The exponent of p in the denominator is

$$B = \left[\frac{m}{p}\right] + \left[\frac{m}{p^2}\right] + \ldots \left[\frac{m}{p^k}\right] + \left[\frac{n}{p}\right] + \left[\frac{n}{p^2}\right] + \ldots +$$

$$+ \ldots + \left[\frac{n}{p^k}\right] + \left[\frac{n+m}{p}\right] + \left[\frac{n+m}{p^2}\right] \ldots + \left[\frac{n+m}{p^k}\right].$$

In order to solve the problem we have to show that $A \geq B$ which follows from

$$\left[\frac{2m}{p^i}\right] + \left[\frac{2n}{p^i}\right] \geq \left[\frac{m}{p^i}\right] + \left[\frac{n}{p^i}\right] + \left[\frac{n+m}{p^i}\right], \quad (i = 1, 2, \ldots, k).$$

This last statement, however, has been proved above at (4), hence the solution is complete.

Remarks. 1. The problem is originally due the Belgian mathematician E. Catalan, who got the result as a byproduct when analysing certain functions.

2. There are various generalizations of the problem; for example, on the Mathematics Olympiads of the United States the following form appeared: Prove that $\dfrac{(5m)!(5m)!}{m!n!(3m+n)!(3n+m)!}$ is an integer.

1972/4. *Find all positive real solutions* $(x_1, x_2, x_3, x_4, x_5)$ *of the system*

(1) $$(x_1^2 - x_3 x_5)(x_2^2 - x_3 x_5) \leq 0,$$

(2) $$(x_2^2 - x_4 x_1)(x_3^2 - x_4 x_1) \leq 0,$$

(3) $$(x_3^2 - x_5 x_2)(x_4^2 - x_5 x_2) \leq 0,$$

(4) $$(x_4^2 - x_1 x_3)(x_5^2 - x_1 x_3) \leq 0,$$

(5) $$(x_5^2 - x_2 x_4)(x_1^2 - x_2 x_4) \leq 0.$$

First solution. Perform the multiplications and sum the left hand sides; their sum will be denoted by B:

$$B = x_1^2 x_2^2 - x_2^2 x_3 x_5 - x_1^2 x_3 x_5 + x_3^2 x_5^2 +$$
$$+ x_2^2 x_3^2 - x_3^2 x_4 x_1 - x_2^2 x_4 x_1 + x_4^2 x_1^2 +$$
$$+ x_3^2 x_4^2 - x_4^2 x_5 x_2 - x_3^2 x_5 x_2 + x_5^2 x_2^2 +$$
$$+ x_4^2 x_5^2 - x_5^2 x_1 x_3 - x_4^2 x_1 x_3 + x_1^2 x_3^2 +$$
$$+ x_5^2 x_1^2 - x_1^2 x_2 x_4 - x_5^2 x_2 x_4 + x_2^2 x_4^2.$$

From this we deduce

$$2B = x_1^2 \left((x_2 - x_4)^2 + (x_3 - x_5)^2 \right) +$$
$$+ x_2^2 \left((x_3 - x_5)^2 + (x_4 - x_1)^2 \right) +$$

$$+ x_3^2 \left((x_4 \perp x_1)^2 + (x_5 \perp x_2)^2 \right) +$$
$$+ x_4^2 \left((x_5 \perp x_2)^2 + (x_1 \perp x_3)^2 \right) +$$
$$+ x_5^2 \left((x_1 \perp x_3)^2 + (x_2 \perp x_4)^2 \right).$$

We conclude that $2B$ is the sum of five nonnegative numbers, therefore $B \geq 0$.

On the other hand, summing the original system gives $B \leq 0$, hence we get that $B = 0$. This, however, can happen only in case all terms in $2B$ are equal to zero:

$$x_2 \perp x_4 = x_3 \perp x_5 = x_4 \perp x_1 = x_5 \perp x_2 = x_1 \perp x_3 = 0,$$

giving $x_1 = x_2 = x_3 = x_4 = x_5$.

The fact that $x_1 = x_2 = x_3 = x_4 = x_5 = c$ for arbitrary positive c satisfies the equality can be seen directly, hence the solutions of the system are given by

$$x_1 = x_2 = x_3 = x_4 = x_5 = c.$$

Second solution. For any positive real c the five tuple

(6) $$\qquad\qquad x_1 = x_2 = x_3 = x_4 = x_5 = c$$

obviously solves the system. Furthermore, if $(x_1, x_2, \ldots, x_5)$ is a solution then so is $\left(\dfrac{1}{x_1}, \dfrac{1}{x_2}, \ldots, \dfrac{1}{x_5} \right)$. Next we show that the assumption of having a solution of the form different from the above ones leads to contradiction. An additional solution violates at least one of the following equations:

$$x_1 = x_3, \quad x_3 = x_5, \quad x_5 = x_2, \quad x_2 = x_4, \quad x_4 = x_1.$$

Since a cyclic permutation of the x_i's leaves the system unchanged, we can assume that

(7) $$\qquad\qquad x_3 < x_5.$$

We divide our argument according to these further solutions satisfy 1) $x_1 \leq \leq x_2$ or 2) $x_1 > x_2$.

1) Let $x_1 \leq x_2$. In this case $x_1^2 \perp x_3 x_5 \leq x_2^2 \perp x_3 x_5$, therefore $x_1^2 \perp x_3 x_5 > > 0$, otherwise $x_2^2 \perp x_3 x_5 > 0$ holds, which contradicts (1). Hence $x_1^2 \perp x_3 x_5 \leq 0$, or equivalently

(8) $$\qquad\qquad x_1 \leq \sqrt{x_3 x_5} < x_5 \quad \text{i.e.} \quad x_1 < x_5.$$

We cannot have $x_2^2 \perp x_3 x_5 < 0$, since it implies $x_1^2 \perp x_3 x_5 < 0$ which again contradicts (1). Consequently, $x_2^2 \perp x_3 x_5 \geq 0$ or

(9) $$\qquad\qquad x_2 \geq \sqrt{x_3 x_5} > x_3, \quad \text{i.e.} \quad x_3 < x_2.$$

(7) and (8) implies

$$x_5^2 > x_1 x_3, \quad \text{i.e.} \quad x_5^2 \perp x_1 x_3 > 0,$$

which — according to (4) — now gives $x_4^2 - x_1x_3 \le 0$, which provides

(10)
$$x_4^2 \le x_1x_3 < x_3x_5$$

by applying (8). On the other hand (7) and (9) shows that $x_3^2 < x_2x_5$, which together with (3) and (9) yields

(11)
$$x_4^2 \ge x_5x_2 > x_5x_3.$$

Now (10) and (11) shows the desired contradiction.

2) Now suppose that $x_1 > x_2$. Then $x_1^2 - x_3x_5 > x_2^2 - x_3x_5$, and so (1) implies that $x_2^2 - x_3x_5$ is nonpositive, hence $x_2^2 \le x_3x_5$ and so

(12)
$$x_2 \le \sqrt{x_3x_5} < x_5.$$

$x_1^2 - x_3x_5 < 0$ implies $x_2^2 - x_3x_5 < 0$ which again contradicts (1), hence $x_1^2 \ge x_3x_5$, or equivalently

(13)
$$x_1 \ge \sqrt{x_3x_5} > x_3.$$

(2) holds only if

$$x_4x_1 \le \max\left(x_2^2, x_3^2\right).$$

Since (12) and (7) imply $x_2^2 \le x_3x_5$ and $x_3^2 \le x_3x_5$, we have

(14)
$$x_4x_1 \le x_3x_5.$$

On the other hand, (5) is satisfied only if

$$x_2x_4 \ge \min\left(x_5^2, x_1^2\right).$$

Since (7) and (13) proves $x_5^2 > x_5x_3$ and $x_1^2 \ge x_5x_3$, we get

(15)
$$x_2x_4 \ge x_5x_3.$$

Comparing (14) and (15) we get $x_4x_1 \le x_2x_4$ or equivalently $x_1 \le x_2$. This, however, contradicts our starting assumption.

This last line now shows that there is no solution different from the ones listed in (6).

1972/5. *Let f and g be two real valued functions defined on the real line satisfying*

(1)
$$f(x+y) + f(x-y) = 2f(x)g(y).$$

Suppose that $f(x)$ is not identically zero and $|f(x)| \le 1$ for all x. Show that $|g(y)| \le 1$ for all y.

First solution. Since the absolute value of a sum does not exceed the sum of absolute values, we have
$$2|f(x)||g(y)| = |f(x+y) + f(x \perp y)| \le |f(x \perp y)| + |f(x+y)|,$$
hence one of the following inequalities hold:
$$|f(x)||g(y)| \le |f(x+y)|, \qquad |f(x)||g(y)| \le |f(x \perp y)|.$$

This means that for y fixed and for every x there is an x_1 (in the form of $x_1 = x + y$ or $x_1 = x \perp y$), satisfying

(2)
$$|f(x)||g(y)| \le |f(x_1)|.$$

Now apply (2) for $x = x_0$; we may assume that $f(x_0) \ne 0$:
$$|f(x_0)||g(y)| \le |f(x_1)|.$$

By iterating the above reasoning we get a sequence $x_0, x_1, \ldots, x_k$ which satisfies
$$|f(x_{k-1})||g(y)| \le |f(x_k)|,$$
hence
$$|f(x_0)||g(y)|^k \le |f(x_k)|,$$
where k is an arbitrary positive integer. Since $|f(x_k)| \le 1$, we have

(3)
$$|g(y)|^k \le \frac{1}{|f(x_0)|}.$$

Now using (3) the existence of a number y with $|g(y)| > 1$ provides a contradiction, since the right hand side is fixed while for suitable k the left hand side can be arbitrarily large.

Second solution. In this solution we will use the fundamental property of reals that every bounded subset of the reals admits a least upper and a greatest lower bound. The least upper bound of a bounded set is well-defined, although it might not belong to the set itself; for example the least upper bound of the real numbers less than 1 is equal to 1.

Since the set of reals $|f(x)|$ is bounded, it admits a least upper bound H, and since $f(x)$ is not identically zero, H is positive; therefore $|f(x)| \le H$ for all x.

Applying the formula for the absolute value of a sum we get
$$2|f(x)||g(y)| \le |f(x+y)| + |f(x \perp y)| \le 2H,$$
hence
(4)
$$|f(x)||g(y)| \le H.$$

Now suppose that there exists y_0 with $|g(y_0)| > 1$. In this case (4) implies that for all x
$$|f(x)| \le \frac{H}{|g(y_0)|} = H_0 < H,$$
showing that the set of values $|f(x)|$ admits an upper bound H_0 strictly less than the least upper bound H. This is, however, a contradiction, and so $|g(y)| \le 1$ is proved for all real y.

Remarks. 1. The condition $|f(x)| \le 1$ was used only in the assumption that $|f(x)|$ is bounded.

2. There are functions satisfying the equation; one example is given by

$$f(x) = \sin x, \quad g(x) = \cos x.$$

1972/6. *Given four parallel planes, prove that there exists a regular tetrahedron with a vertex on each plane.*

First solution. Suppose that the tetrahedron $ABCD$ exists with vertices A, B, C, D on the parallel planes α, β, γ, δ. The planes are on one side of α and their distances from α are b, c and d, respectively. On *Figure 1972/6.1* these planes are perpendicular to the plane of the figure, hence they appear as parallel lines.

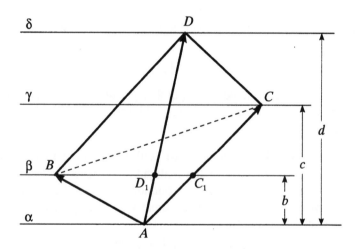

Figure 72/6.1

β intersects AC in C_1 and AD in D_1. Since the ratio of the distances between α—β and α—γ is $b : c$, the same holds for the ratio of AC_1 and AC:

(1)
$$\frac{AC}{AC_1} = \frac{c}{b}.$$

For the same reason

(2)
$$\frac{AD}{AD_1} = \frac{d}{b}.$$

Notice that an enlargement with centre A transforms our regular tetrahedron into another regular tetrahedron. The planes passing through the vertices are mapped under this enlargement into planes with the same ratios of distances.

Based on the above observations the existence of the desired solid can be verified as follows: consider an arbitrary regular tetrahedron $ABCD$. Fix C_1 and D_1 on AC and AD such that they satisfy (1) and (2). Since B, C_1 and D_1 are not collinear, they determine a plane β'. Apply now an enlargement with centre A transforming the distance of A and β' to b and mapping the plane α passing through A and parallel to β'. After the enlargement the distances of C and D from α become c and d, therefore the parallel planes to α passing through B, C and D are of distances b, c and d from α. This shows that the desired tetrahedron exists.

Second solution. The original version of the problem received by the jury at the competition also asked the edge length of the tetrahedron in terms of the distances of the parallel planes. The following analytic geometric solution will answer this question as well.

Let us choose our coordinate system in such a way that the coordinates of the vertices of the tetrahedron become $A(0,0,0)$, $B(3s, 3s\sqrt{3}, 0)$, $C(\perp 3s, 3s\sqrt{3}, 0)$, $D(0, 2s\sqrt{3}, 2s\sqrt{6})$ (the edge length of this tetrahedron is $6s$). The unit vector $\mathbf{n}$ starting at A has coordinates $\mathbf{n}(p, q, r)$, so

(3) $$p^2 + q^2 + r^2 = 1.$$

We will use the notations $\overrightarrow{AB} = \mathbf{b}$, $\overrightarrow{AC} = \mathbf{c}$, $\overrightarrow{AD} = \mathbf{d}$. The orthogonal projections of B, C, D to the line of $\mathbf{n}$ are denoted by B', C', D', respectively. If we can choose $\mathbf{n}$ and s in a way that $AB' = b$, $AC' = c$, $AD' = d$ is satisfied (where b, c, d are given positive distances) then the planes α, β, γ, δ orthogonal to $\mathbf{n}$ at the points A, B, C, D will have the above prescribed distances and so the points provide the desired regular tetrahedron $ABCD$.

Now the coordinates of the vectors representing the edges are as follows:

$$\mathbf{b}(3s, 3s\sqrt{3}, 0) \quad \mathbf{c}(\perp 3s, 3s\sqrt{3}, 0) \quad \mathbf{d}(0, 2s\sqrt{3}, 2s\sqrt{6}).$$

Since the scalar product with a unit vector is equal to the (signed) length of the projection to the line determined by the unit vector, we have

$$\mathbf{nb} = b, \quad \mathbf{nc} = c, \quad \mathbf{nd} = d,$$

or in coordinates

$$3sp + 3\sqrt{3}sq = b,$$
$$\perp 3sp + 3\sqrt{3}sq = c,$$
$$sq + 2\sqrt{6}sr = d.$$

This system can be easily solved for the unknowns sp, sq, sr and we get:

(4) $$sp = \frac{b \perp c}{6}, \quad sq = \frac{b+c}{6\sqrt{3}}, \quad sr = \frac{3d \perp b \perp c}{6\sqrt{6}}.$$

In view of (3) this gives

$$(sp)^2 + (sq)^2 + (sr)^2 = s^2 = \frac{1}{36 \cdot 6}\left(\left(\sqrt{6}(b \perp c)\right)^2 + \left(\sqrt{2}(b+c)\right)^2 + (3d \perp b \perp c)^2\right) =$$

$$= \frac{1}{36 \cdot 2} \left(b^2 + c^2 + d^2 + (b \perp c)^2 + (c \perp d)^2 + (d \perp b)^2 \right).$$

In conclusion, the edge length of the tetrahedron is equal to

(5) $$6s = \frac{1}{\sqrt{2}} \sqrt{b^2 + c^2 + d^2 + (b \perp c)^2 + (c \perp d)^2 + (d \perp b)^2}.$$

(4) also shows the coordinates of the unit vector **n**:

(6) $$p = \frac{b \perp c}{6s}, \quad q = \frac{b+c}{6s\sqrt{3}}, \quad r = \frac{3d \perp b \perp c}{6s\sqrt{6}}.$$

Consider a tetrahedron $ABCD$ of edge length $6s$ given by (5) in the coordinate system and fix the unit vector **n** with coordinates (p, q, r) given by (6). The planes α, β, γ and δ orthogonal to **n** passing through the vertices of this tetrahedron will have the prescribed distances; consequently the desired tetrahedron exists.

Remark. Notice that the first solution did not use the fact that the tetrahedron is regular, consequently it applies to any tetrahedron.

1973.

1973/1. *Let $\overrightarrow{OP_1}, \overrightarrow{OP_2}, \ldots, \overrightarrow{OP_n}$ be unit vectors in a plane. $P_1, P_2, \ldots, P_n$ all lie on the same side of a line through O. Prove that if n is odd, then*

$$|\overrightarrow{OP_1} + \overrightarrow{OP_2} + \ldots + \overrightarrow{OP_n}| \geq 1,$$

where $|\overrightarrow{OM}|$ denotes the length of the vector $\overrightarrow{OM}$.

First solution. First, observe that if the angle of the vectors **a** and **b** is at most 90°, then

(1) $$|\mathbf{a}+\mathbf{b}| \geq |\mathbf{a}| \quad \text{and} \quad |\mathbf{a}+\mathbf{b}| \geq |\mathbf{b}|.$$

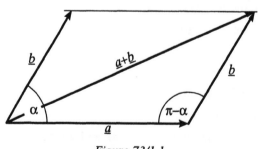

Indeed, if **a**, **b**, **a**+**b** are the sides of a triangle where the angle opposite to **a**+**b** is at most 90°, then it is the largest angle of the triangle, so the longest side is $|\mathbf{a}+\mathbf{b}|$. (1) holds even if one of **a** or **b** is the nil-vector, in this case their angle is 0° (see *Figure 1973/1.1*).

Figure 73/1.1

Let A and B denote the point of intersection of the line e and the unit circle with centre O, and order the points P_i from A to B. So the points lie on the arc between P_1 and P_n (see *Figure 1973/1.2*).

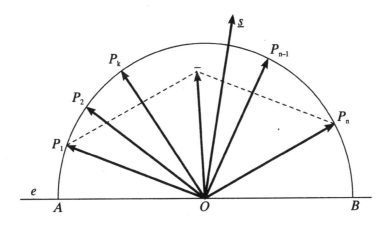

Figure 73/1.2

We extend the statement of the problem with the following condition: the sum of the vectors lies in the angular domain P_1OP_n. We prove the statement by induction: for $n=1$ the statement is obvious; assume that it is true for $n \perp 2$, that is, for the vector $s = \overrightarrow{OP_2} + \overrightarrow{OP_3} + \ldots + \overrightarrow{OP_{n-1}}$

$$|s| = |\overrightarrow{OP_2} + \overrightarrow{OP_3} + \ldots + \overrightarrow{OP_{n-1}}| \geq 1,$$

and s lies in the angle P_2OP_{n-1}. The vector $\overrightarrow{OP_1} + \overrightarrow{OP_n} = v$ lies along the bisector of the angle P_1OP_n, and so by our introductory remark it makes an acute angle with s. Hence:

$$|v+s| = |\overrightarrow{OP_1} + \overrightarrow{OP_2} + \ldots + \overrightarrow{OP_n}| \geq |s| \geq 1.$$

It is easy to see that $v+s$ lies in the angular domain P_1OP_n, and this is what we wanted to prove.

Second solution. Using the notation of the first solution, denote $\overrightarrow{OP_i}$ by e_i ($i=1, 2, \ldots, n$). For arbitrary $1 \leq i, j \leq n$ the vector $e_j + e_i$ bisects the angle of e_j and e_i, and so $e_j + e_i$ makes an acute angle with every vector e_s in the angular domain bounded by e_j and e_i. Thus $e_s(e_j + e_i) \geq 0$. If $n = 2k \perp 1$, the "middle" vector, e_k is in the angles determined by the pairs of vectors (e_1, e_{2k-1}), (e_2, e_{2k-2}), $\ldots$, (e_{k-1}, e_{k+1}). So:

$$e_k(e_1 + e_2 + \ldots + e_{k-1} + e_k + e_{k+1} + \ldots + e_{2k-2} + e_{2k-1}) =$$

(2) $= e_k(e_1 + e_{2k-1}) + e_k(e_2 + e_{2k-2}) + \ldots + e_k(e_{k-1} + e_{k+1}) + e_k^2 \geq e_k^2 = 1.$

On the other hand,

(3) $e_k(e_1 + e_2 + \ldots + e_n) \leq |e_k||e_1 + e_2 + \ldots + e_n| = |e_1 + e_2 + \ldots + e_n|.$

Now (2) and (3) gives

$$|e_1 + e_2 + \ldots + e_n| \geq 1.$$

1973/2. *Can we find a finite set of non-coplanar points, such that given any two points, A and B, there are two others, C and D, with the lines AB and CD parallel and distinct?*

Solution. Let $ABCD$ and $A'B'C'D'$ the bottom and the top faces of a cube. Let the reflection of S, the centre of the cube, to the bottom and the top faces K and K', respectively. The 8 vertices of the cube along with K and K' make a set of points, M satisfying the conditions of the problem. (See *Figure 1973/2.1*).

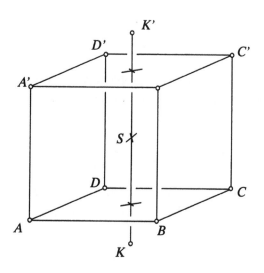

Figure 73/2.1

Indeed, pick two points of M. If S does not lie on the line through them, then it is parallel to its reflection to S. This reflection is through two points of M, too, as M is symmetric to S. If this line is a main diagonal of the cube, then it is parallel to one of the segments KA, KB, KC, KD. Finally, KK' is parallel, for example, to AA'.

Remark. There are several ways to construct a set of points M, satisfying the conditions of the problem. Our example can be generalized in two ways: Instead of a cube we can take any regular prism with even many faces. Alternatively, our example can be realized as the union of two central symmetric hexagons, $AA'K'C'CK$ and $BB'K'D'DK$ with the common diagonal KK'. Making a similar construction with e.g. two regular $2n$-gons the resulting $4n \perp 2$ points satisfy the conditions of the problem.

The 12 midpoints of the edges of a cube is a good construction, as well. For a non central symmetric M put together 5 cubes of the same size, such that their centres are collinear and the neighbour cubes have a face in common. The 24 vertices and the centre of the first, second and fifth cubes form a set of 27 points that satisfies the conditions.

1973/3. *Let a and b be real numbers for which the equation $x^4 + ax^3 + bx^2 + ax + 1 = 0$ has at least one real solution. Find the least possible value of $a^2 + b^2$.*

First solution. As $x = 0$ is not the root of the equation, we can divide by x^2. After reordering we get

$$x^2 + \frac{1}{x^2} + a\left(x + \frac{1}{x}\right) + b = 0.$$

Let $x + \dfrac{1}{x} = z$. Thus $x^2 + \dfrac{1}{x^2} = z^2 \perp 2$. The sum of a positive number and its reciprocal is at least 2, so

(1) $$z^2 = x^2 + \frac{1}{x^2} + 2 \geq 4.$$

Using the substitutions above, we get

(2) $$z^2 + az + b \perp 2 = 0,$$
$$2 \perp z^2 = az + b.$$

Applying the Cauchy-inequality ([22]) to the right hand side of the equation:

(3) $$(2 \perp z^2)^2 = (a \cdot z + b \cdot 1)^2 \leq (a^2 + b^2)(z^2 + 1),$$

yielding

(4) $$a^2 + b^2 \geq \frac{(2 \perp z^2)^2}{z^2 + 1} = (z^2 \perp 2)\left(1 \perp \frac{3}{z^2 + 1}\right).$$

Both factors of the product on the right hand side is increasing when z^2 is increasing, so according to (1) it takes its smallest value at $z^2 = 4$:

$$a^2 + b^2 \geq \frac{4}{5}.$$

Hence the minimum of $a^2 + b^2$ is $\dfrac{4}{5}$; we have to show that under these conditions the original equation has a real root. Since in this case there is an equality in (3), by the Cauchy inequality

$$a:b=z:1, \quad \text{that is} \quad a^2=b^2z^2,$$

must hold, and so $a^2=4b^2$. Hence

$$a^2+b^2=5b^2=\frac{4}{5}, \quad b^2=\frac{4}{25}, \quad a^2=\frac{16}{25}.$$

The possible values of the coefficients are:

$$a=\pm\frac{4}{5}, \quad b=\pm\frac{2}{5}.$$

If we choose the negative value in both cases, $x=1$ is the root of

$$x^4-\frac{4}{5}x^3-\frac{2}{5}x^2-\frac{4}{5}x+1=0,$$

and so the least possible value of a^2+b^2 is $\frac{4}{5}$.

Second solution. Let us assume that x is a positive solution of the equation. As for every real c the inequality $c\geq-|c|$ holds,

$$0=x^4+ax^3+bx^2+ax+1\geq x^4-|a|x^3-|b|x^2-|a|x+1,$$

hence

(5) $$x^4+1\leq|a|x^3+|b|x^2+|a|x=|a|(x^3+x)+\frac{|b|}{2}2x^2.$$

Now, as $(x^2-1)^2\geq 0$, we have

$$2x^2\leq x^4+1.$$

Since the sign of x^3-1 and $x-1$ are the same, $(x^3-1)(x-1)\geq 0$ follows, and so

$$x^3+x\leq x^4+1;$$

from (5) we get

$$x^4+1\leq|a|(x^4+1)+\frac{|b|}{2}(x^4+1),$$

that is

$$1\leq|a|+\frac{|b|}{2},$$

$$|b|\geq 2-2|a|=2(1-|a|).$$

If $|a|\leq 1$,

$$a^2+b^2=|a|^2+|b|^2\geq|a|^2+4+4|a|^2-8|a|=5\left(|a|-\frac{4}{5}\right)^2+\frac{4}{5},$$

thus

(6) $$a^2+b^2\geq\frac{4}{5}$$

and equality can only hold if $|a|=\frac{4}{5}$.

If $|a| > 1$, then (6) clearly holds.

Now, if we assume that the original equation has a negative solution, then the positive x is a solution of the equation

$$x^4 \perp ax^3 + bx^2 \perp ax + 1 = 0;$$

and, again we get (6). As in the first solution, we can choose the coefficients such that $a^2 + b^2 = \frac{4}{5}$ and the equation has a real root. Hence the minimum of $a^2 + b^2$ is $\frac{4}{5}$

Remark. The equation in the problem is a so called reciprocal equation; a polynomial equation is called reciprocal if along with the root x_0 the value $\frac{1}{x_0}$, (or $\perp \frac{1}{x_0}$) is a root, too. The coefficients of these equations are symmetric (or antisymmetric), i.e. the $n \perp k$-th and the k-th coefficients of the polynomial of degree n are equal (or opposites). The substitutions from the first solution work even for reciprocal polynomials of degree 9.

1973/4. *A soldier needs to sweep a region of the shape of an equilateral triangle for mines. The detector has an effective radius equal to half the altitude of the triangle. He starts at a vertex of the triangle. What path should he follow in order to travel the least distance and still sweep the whole region?*

Solution. Let us consider the triangle ABC, with side 1, and suppose that the soldier starts at A. In order to sweep the other two vertices, his path must intersect the circles k_1 and k_2 of radius $\frac{m}{2}$ with centre B and C, respectively. By symmetry we may assume that he first intersects k_1 at the point P. The shortest path from P to k_2 lies on the line PC. Let Q denote the point of intersection of k_2 and PC. The path $AP + PC$ is longer than $AP + PQ$ by $\frac{m}{2}$, hence $AP + PC$ is the shortest when $AP + PQ$ is the shortest (see Figure 1973/4.1).

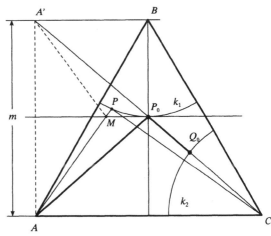

Figure 73/4.1

First, we are looking for the point P_0 on the circle k_1, such that $AP_0 + P_0C$ is the shortest. We show that P_0 is the point where the altitude of B intersects the circle. Let e be the tangent of k_1 at P_0 and pick an arbitrary point P distinct from P_0 on k_1. We prove that $AP + PC > AP_0 + P_0C$. Let A' denote the reflection of A to e. Since $AA' = m$, the points A', P_0, C are collinear and so $AP_0 + P_0C = = A'C$. The line e separates k_1 from A thus the point of intersection of AP and e is a single point M. Clearly, $AM = A'M$, so $AP + PC = A'M + MP + PC$; the polygonial path $A'MPC$ is longer than $A'C$, thus

$$AP + PC < AP_0 + P_0C.$$

Now, if Q_0 denotes the intersection of k_2 and P_0, the shortest path is: $A \rightarrow P_0 \rightarrow \rightarrow Q_0$.

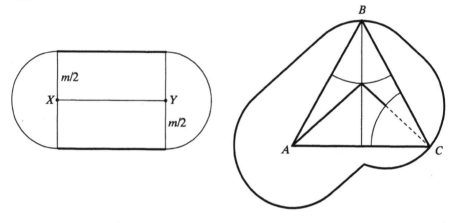

Figure 73/4.2   *Figure 73/4.3*

It remains to check that the whole triangle can be swept from this path. For a segment XY the points not farther than $\dfrac{m}{2}$ cover a "stadium-shaped region" that is shown of *Figure 1973/4.2*. The stadium-regions belonging to AP_0 and P_0Q_0 cover the ABC triangle, so the conditions are satisfied by the $A \rightarrow P_0 \rightarrow \rightarrow Q_0$ path. (See *Figure 1973/4.3*).

1973/5. *Let G be a set of non-constant functions f. Each f is defined on the real line and has the form $f(x) = ax + b$ for some real a, b.*

a) *If $f, g \in G$, then so is $g \circ f \in G$, where $g \circ f(x) = g(f(x))$.*

b) *If f is in G, then so is the inverse f^{-1}.*

$$f^{-1}(x) = \frac{x}{a} \perp \frac{b}{a}.$$

c) *Every f in G has a fixed point. In other words we can find x_f such that*

$$f(x_f) = x_f.$$

Prove that all the functions in G have a common fixed point.

Solution. f not being constant means that $a \neq 0$. From c) we get that if in $f(x) = ax + b$, the condition $a = 1$ implies $b = 0$, because $x_f + b = x_f$ cannot be true.

From b) $f^{-1}(x) = \dfrac{x}{a} \perp \dfrac{b}{a}$ and from a)

$$f \circ f^{-1} = f \circ f^{-1}(x) = a\left(\frac{x}{a} \perp \frac{b}{a}\right) + b = x$$

belong to G. Set $e(x) = x$; as G is non-empty, $e(x)$ is in G; every real x is a fixpoint of $e(x)$.

If G contains only $e(x)$, the statement is obvious. If G has an element f not equal to e, then it has infinitely many elements as $f, f \circ f, (f \circ f) \circ f, \ldots$ are all distinct functions, because the coefficients of x are $a, a^2, a^3, \ldots$ and they are all distinct.

Now, let $f(x) = ax + b$, $g(x) = cx + d$ $(a \neq 1, c \neq 1)$ be two arbitrary elements of G. We have to show that if k is a fixpoint of f, then it is a fixpoint of g, as well. If k is a fixpoint of f then

$$ak + b = k, \quad \text{that is} \quad k = \frac{b}{1 \perp a}.$$

Similarly, the fixpoint of g is $\dfrac{d}{1 \perp c}$, hence to prove our statement we have to show that

(1) $$k = \frac{b}{1 \perp a} = \frac{d}{1 \perp c}.$$

Consider the functions $p = f \circ g$ and $q = g \circ f$:

$$p(x) = a(cx + d) + b = acx + ad + b$$
$$q(x) = c(ax + b) + d = acx + bc + d.$$

The composition $p \circ q^{-1}$ is an element of G, too:

$$p \circ q^{-1}(x) = ac\left(\frac{x}{ac} \perp \frac{bc + d}{ac}\right) + ad + b = x + (\perp bc \perp d + ad + b).$$

Since the coefficient of x equals 1, $\perp bc \perp d + ad + b = 0$, hence

$$b(1 \perp c) = d(1 \perp a)$$

$$\frac{b}{1 \perp a} = \frac{d}{1 \perp c}$$

and (1) really holds.

Remarks. 1. In this problem, the set G is a group ([23]). As for $e(x) = x$ we have $f \circ e = e \circ f = f$, e is the unit element of the group; in order to show that G is a group, it only remained to prove that G is associative, i.e if $f, g, h \in G$, then $(f \circ g) \circ h = f \circ (g \circ h)$ which is an easy calculation.

G is called a group of linear mappings of the line. As we showed, for every $f(x) = ax + b$ in G, $f(k) = k$, that is $ak + b = k$, $b = k(1 \perp a)$; hence every element of G is of the form

(2) $$f(x) = ax + k(1 \perp a).$$

2. The following observations show the geometric motivation of the problem: (2) implies that f is a similarity transformation of the line with centre k

and ratio a. Indeed, let x be an arbitrary point of the real line. Its distance from k is $x \perp k$, the the distance of $f(x)$ from k is $f(x) \perp k$, reordering (2) we get

$$\frac{f(x) \perp k}{x \perp k} = a,$$

which means that the proportion of the distances of $f(x)$ and x equal to a constant a. The geometrical meaning of c) says that there is no translation in G, as every element has a fixed point. The statement itself is a well known theorem in geometry: If a set of similarities of the line form a group that contains no translations, then the elements of the set are central similarities with a given centre (The theorem holds in higher dimensions, too).

Even the proof of the theorem has some geometric motivation based on the following: Apply enlargements with proportions a and then c, followed by enlargements with proportions $\frac{1}{a}$ and $\frac{1}{c}$ to the plane then the resulting objects do not change size and direction of the objects. As there is no translation, it has to be the identity transformation.

2. An other geometric correspondence is the following one: G is a set of lines with equation $y = ax + b$ that are not parallel to the x axes and have the following properties:

a) If the lines $y = ax + b$ and $y = cx + d$ belong to G then so does $y = acx + (ad + b)$.

b) If a line is in G, then so is its reflection to $y = x$.

c) Every line in G meets the $y = x$ line.

Corollary: Every element of G contains a point $K(k, k)$ in common.

1973/6. *Let $a_1, a_2, \ldots, a_n$ be positive reals, and q satisfies $0 < q < 1$. Find $b_1, b_2, \ldots, b_n$ such that:*

a) $a_k < b_k$, $(k = 1, 2, \ldots, n)$;

b) $q < \dfrac{b_{k+1}}{b_k} < \dfrac{1}{q}$, $(k = 1, 2, \ldots, n \perp 1)$;

c) $b_1 + b_2 + \ldots + b_n < \dfrac{1+q}{1 \perp q} (a_1 + a_2 + \ldots + a_n)$.

Solution. Condition c) suggests us to construct the numbers b_k in the following way:

$$b_1 = a_1 + a_2 q + a_3 q^2 + \ldots + a_n q^{n-1}$$

$$b_2 = a_1 q + a_2 + a_3 q + \ldots + a_n q^{n-2}$$

(1)
$$b_3 = a_1 q^2 + a_2 q + a_3 + \ldots + a_n q^{n-3}$$

$$\cdots\cdots\cdots\cdots\cdots\cdots\cdots$$

$$b_n = a_1 q^{n-1} + a_2 q^{n-2} + a_3 q^{n-3} + \ldots + a_n.$$

In general

(2)
$$b_k = \sum_{i=1}^{k-1} a_i q^{k-i} + a_k + \sum_{i=k+1}^{n} a_i q^{i-k}.$$

We show that the construction in (1) satisfies conditions a)–c). a) is obvious as on the right hand side of (1) there are additional positive summands with a_k. for b) observe that

$$b_{k+1} = \sum_{i=1}^{k} a_i q^{k+1-i} + \sum_{i=k+1}^{n} a_i q^{i-k-1} = q \sum_{i=1}^{k} a_i q^{k-1} + \frac{1}{q} \sum_{i=k+1}^{n} a_i q^{i-k}.$$

As $0 < q < 1$ and $\dfrac{1}{q} > 1$, we conclude

$$b_{k+1} < \frac{1}{q} \sum_{i=1}^{k} a_i q^{k-i} + \frac{1}{q} \sum_{i=k+1}^{n} a_i q^{i-k} = \frac{1}{q} b_k,$$

$$b_{k+1} > q \sum_{i=1}^{k} a_i q^{k-i} + q \sum_{i=k+1}^{n} a_i q^{i-k} = q b_k,$$

Thus we proved both inequalities in b).

To get c) sum the the columns of the expressions in (1) ; the sum of the i-th column:

$$a_i \left(q^{i-1} + q^{i-2} + \ldots + q + 1 \right) + a_i \left(q + q^2 + \ldots + q^{n-i} \right) =$$

$$= a_i \left(\frac{1 \perp q^i}{1 \perp q} + q \frac{1 \perp q^{n-i}}{1 \perp q} \right) = a_i \frac{1 + q \perp q^i \perp q^{n-i+1}}{1 \perp q} < a_i \frac{1+q}{1 \perp q}.$$

Summing up the n sums of the columns:

$$b_1 + b_2 + \ldots + b_n < (a_1 + a_2 + \ldots + a_n) \frac{1+q}{1 \perp q},$$

hence c) is satisfied.

Remark. Our proof shows that the a_i-s do not uniquely determine the b_i-s. Indeed, there are infinitely many n-tuples for b_i-s that satisfy the conditions.

1974.

1974/1. *Three players A, B, C play the following game: There are three cards each with a different positive integer p, q and r where $p < q < r$. In each round the cards are randomly dealt to the players and each receives the number of counters on his card.*

After two or more rounds, A has received 20, B 10 and C 9 counters. In the last round B received the largest number of counters.

Who received q counters in the first round?

Solution. Let us assume that they played n rounds $(n \geq 2)$. In each round the total of the cards is $p+q+r$, so the number of counters received altogether is $n(p+q+r)$. By the assumptions this equals to $20+10+9=39$:

$$(1) \qquad\qquad n(p+q+r)=3 \cdot 13.$$

Since $p \geq 1$, $q \geq 2$, $r \geq 3$ and $n \geq 2$, $p+q+r \geq 6$, and (1) can only hold if $n=3$ and $p+q+r=13$. The highest score was 20, so r cannot be smaller than 7, so $r \geq 7$. As $p+q+r > 10$ and B received r counters in the last round and altogether 10 counters, the distribution of B's counters cannot be $q+q+r$, $p+$ $+q+r$, and could not receive r counters twice. So the number of B's counters is $p+p+r$, so

$$2p+r=10,$$

and $p \geq 1$, $r \geq 7$ implies $p=1$, $r=8$ and so $q=4$.

C had 9 counters, so he could not get $r=8$ counters, only 1 or 4. As B received the card with 1 in the first round, C had to receive the card 4. Thus in the first round C received the $q=4$ counters. Now, we can set the table of the game:

	A	B	C
	number of counters		
I. round	8	1	4
II. round	8	1	4
III. round	4	8	1
All	20	10	9

Remark. The following question was raised at the evaluation of the solutions: does one have to make the table, i.e. verify the existence of the game, or was it assumed. (They accepted the latter one). If we do not assume the existence of the game, we can derive a wrong conclusion. For example, there is no game that can result 18, 8 and 7 counters, respectively, although even in this case our method shows that C received the q counters in the first round. The game (and so the table) does not exist, though.

1974/2. *Let A, B and C denote the vertices of a triangle. Prove that there is a point D on the side AB of the triangle ABC, such that CD is the geometric mean of AD and DB if and only if*

(1) $$\sin A \sin B \le \sin^2 \frac{C}{2}.$$

First solution. We have to show that (1) holds if and only if there is a point D on AB such that

(2) $$AD \cdot DB = CD^2.$$

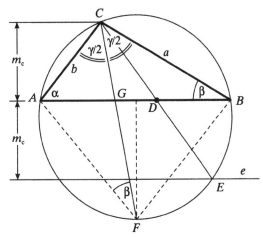

Let us assume that such a D exists and CD intersects the circumcircle of the triangle in E (see *Figure 1974/2.1*). The product of the lengths of the segments through an inner point of the circle is constant, so $AD \cdot DB = CD \cdot DE$, hence from (2) we have $CD = DE$, that is the distance of E from AB equals the altitude at C. Thus the point D exists, if and only if the line e parallel to AB of distance h_c from AB intersects the circumcircle. This intersection exists, if the distance

Figure 74/2.1

of F, the midpoint of the AB arc not containing C from AB is at most h_c. This holds if and only if the area of the triangle AFB is at least the area of the triangle ABC: $F_{AFB} \ge F_{ABC}$.

Let the length of the diameter of the circumcircle equal to 1. Then, $BC = \sin A$, $AC = \sin B$ and so

$$F_{ABC} = \frac{1}{2} \sin A \sin B \sin C.$$

In the isosceles triangle, AFB we have $\angle AFB = 180° \perp C$, the other two angles equal to $\frac{C}{2}$, so its area

$$F_{AFB} = \frac{1}{2} \sin^2 \frac{C}{2} \sin C,$$

and the inequality for the areas is equivalent to

$$\frac{1}{2} \sin A \sin B \sin C \le \frac{1}{2} \sin^2 \frac{C}{2} \sin C.$$

that is equivalent to (1). Thus we proved the equivalence of (1) and (2).

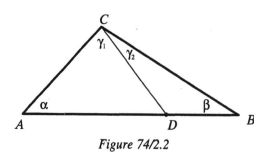

Figure 74/2.2

Second solution. Let us assume, again, the existence of the point D, and suppose that the angle C is split into C_1, and C_2 by CD (see *Figure 1974/2.2*). Apply the Law of Sines to the triangles ADC, and BDC:

(3) $\quad \dfrac{\sin A}{\sin C_1} = \dfrac{CD}{AD}, \quad \dfrac{\sin B}{\sin C_2} = \dfrac{CD}{DB}.$

Multiplying the two equations and applying (2) we get:

(4) $$\frac{\sin A \sin B}{\sin C_1 \sin C_2} = \frac{CD^2}{AD \cdot DB} = 1,$$

(5) $$\sin A \sin B = \sin C_1 \sin C_2.$$

As $\ 2 \sin C_1 \sin C_2 = \cos(C_1 - C_2) - \cos(C_1 + C_2)\ $ and $\ \cos C = 1 - 2\sin^2 \dfrac{C}{2},\ $ we conclude

$$2 \sin A \sin B = \cos(C_1 - C_2) - \cos C = 2\sin^2 \frac{C}{2} - (1 - \cos(C_1 - C_2)) \le 2\sin^2 \frac{C}{2},$$

so we showed that (2) implies (1).

Now, if (1) holds, from $2\sin^2 \dfrac{C}{2} = 1 - \cos C$ it follows that

$$2\sin A \sin B \le 1 - \cos C, \qquad 2\sin A \sin B + \cos C \le 1.$$

Since $0 < 2\sin A \sin B$ and $-1 < \cos C$,

$$-1 < 2\sin A \sin B + \cos C \le 1.$$

This means that there is an angle φ such that $0 \le \varphi < \pi$ and the following holds:

(6) $$\cos \varphi = 2\sin A \sin B + \cos C.$$

As $\cos \varphi > \cos C$ we have $0 \le \varphi < C$, introduce the following notations: $C_1 = \dfrac{C + \varphi}{2}$, $C_2 = \dfrac{C - \varphi}{2}$. From (6)

$$2\sin A \sin B = \cos\varphi - \cos C = 2 \sin \frac{C + \varphi}{2} \sin \frac{C - \varphi}{2} = 2 \sin C_1 \sin C_2.$$

This means that if we split C into C_1 and C_2 then (5) holds, and hence (3) implies (4) and so (2), as well. Thus (1) implies (2) and this is what we wanted to prove.

Third solution. We prove that both (1) and (2) are equivalent to the following inequality for the sides of the triangle:

$$a + b \le c\sqrt{2}.$$

Let the bisector of C intersect the circumcircle in F, and the side AB in G. By the remark referring to *Figure 1974/2.1* in the first solution the point D exists if and only if CF intersects e, i.e. $GF \ge CG$.

The triangles CGB and CAF are similar, because two of their angles are measured by the same arcs. From the equality of the proportion of the sides we have:

$$\frac{a}{CG} = \frac{CF}{b} = \frac{CG + GF}{b} \qquad \text{that is} \qquad CG^2 = ab \perp CG \cdot GF$$

As $GF \geq CG$ and the equality of the products of the segments of a chord $CG \cdot$
$\cdot GF = AG \cdot GB$, we obtain

$$AG \cdot GB = CG \cdot GF \geq CG^2 = ab \perp AG \cdot GB,$$

(7) $$2AG \cdot GB \geq ab.$$

Also, the bisector splits the opposite side into the proportion of the two adjacent sides, hence we get:

$$AG = \frac{bc}{a+b}, \qquad GB = \frac{ac}{a+b},$$

Substituting this into (7) we obtain:

$$\frac{2abc^2}{(a+b)^2} \geq ab,$$

and

(8) $$a + b \leq c\sqrt{2}.$$

As the reverse of our arguments are true, (8) is equivalent to (2).

Now, we prove that (1) and (8) are equivalent, as well. Let us assume that (1) holds and choose the diameter of the circumcircle to be equal to 1. The Law of Cosine $(2ab(1 + \cos C) = (a+b)^2 \perp c^2)$ implies:

$$2 \sin A \sin B \leq 2 \sin^2 \frac{C}{2} = \frac{4 \sin^2 \frac{C}{2} \cos^2 \frac{C}{2}}{2 \cos^2 \frac{C}{2}} = \frac{\sin^2 C}{2 \cos^2 \frac{C}{2}} = \frac{\sin^2 C}{1 + \cos C}.$$

Substituting $a = \sin A$, $b = \sin B$, $c = \sin C$ we get

$$2ab \leq \frac{c^2}{1 + \cos C}, \qquad 2ab(1 + \cos C) = (a+b)^2 \perp c^2 \leq c^2,$$

and (8) can easily derived from here. The reverse of our arguments hold, so (1) and (8) are equivalent.

Remark. This last result shows that the point D exists if $a+b$ is not "very far" from c, since $c < a+b \leq c\sqrt{2} \approx 1{,}41c$. This always holds if C is not acute, because in this case from the Law of Cosine $c^2 \geq a^2 + b^2$, and so

$$2c^2 \geq 2(a^2 + b^2) = (a+b)^2 + (a \perp b)^2 \geq (a+b)^2.$$

1974/3. *Prove that*

$$\sum_{k=0}^{n} \binom{2n+1}{2k+1} \cdot 2^{3k}$$

is not divisible by 5 for any nonnegative integer n.

First solution. The sum in the problem reminds us to the binomial theorem. As $\sqrt{2^3} = \sqrt{8}$, by this theorem

$$(\sqrt{8}+1)^{2n+1} = \left(2\sqrt{2}+1\right)^{2n+1} = \sum_{k=0}^{2n+1} \binom{2n+1}{k}(\sqrt{8})^k =$$

$$= \sum_{k=0}^{n} \binom{2n+1}{2k} \cdot 8^k + \sum_{k=0}^{n} \binom{2n+1}{2k+1} \cdot 8^k \cdot \sqrt{8}.$$

Set

$$a_n = \sum_{k=0}^{n} \binom{2n+1}{2k} \cdot 8^k, \quad b_n = \sum_{k=0}^{n} \binom{2n+1}{2k+1} \cdot 8^k.$$

So b_n denotes our sum, and

(1) $$(\sqrt{8}+1)^{2n+1} = a_n + b_n\sqrt{8}.$$

Similarly, we get

(2) $$(\sqrt{8}\perp 1)^{2n+1} = \perp a_n + b_n\sqrt{8}.$$

Multiplying the two equalities we conclude:

$$7^{2n+1} = 8b_n^2 \perp a_n^2,$$

(3) $$7 \cdot 49^n + a_n^2 = 8b_n^2.$$

As a_n and b_n are integers, in case 5 divided b_n the last digit of the right hand side of (3) would equal to 0. The last digit of 49^n is 9 or 1, hence the last digit of $7 \cdot 49^n$ is 3 or 7. The possible last digits of a_n^2 are 0, 1, 4, 5, 6, 9, the last digit of the left hand side cannot equal 0, so 5 does not divide b_n.

Second solution. We start with formula (1):

$$a_{n+1} + b_{n+1}\sqrt{8} = (\sqrt{8}+1)^{2(n+1)+1} = \left(\sqrt{8}+1\right)^{2n+1}\left(\sqrt{8}+1\right)^2 =$$

$$= (a_n + b_n\sqrt{8})(9+2\sqrt{8}) = (9a_n + 16b_n) + (2a_n + 9b_n)\sqrt{8}.$$

The numbers in the parenthesis are integers, so

$$a_{n+1} = 9a_n + 16b_n \qquad b_{n+1} = 2a_n + 9b_n.$$

As we are interested in the divisibility by 5, from now on we handle these equations as congruences mod 5

(4) $$a_{n+1} \equiv 4a_n + b_n, \qquad b_{n+1} \equiv 2a_n + 4b_n.$$

Repeating the procedure:

$$a_{n+2} \equiv 4a_{n+1} + b_{n+1} \equiv 3a_n + 3b_n,$$
$$b_{n+2} \equiv 2a_{n+1} + 4b_{n+1} \equiv a_n + 3b_n.$$

Finally

$$b_{n+3} \equiv 2a_{n+2} + 4b_{n+2} \equiv \quad 3b_n.$$

Thus in the series b_n an element is divisible by 5 if and only if the third preceding element is divisible by 5. From (1) $a_0 = b_0 = 1$ and from (4) $a_1 \equiv 0$,

$b_1 \equiv 1$; $b_2 \equiv 4$, so 5 does not divide b_0, b_1, and b_2, hence there is no b_n divisible by 5.

1974/4. *An 8×8 chessboard is divided into p disjoint rectangles (along the lines between squares), so that*

a) *each rectangle has the same number of white squares as black squares*

b) *If a_i denotes the number of white squares in the i-th rectangle, then $a_1 <$ $< a_2 < \ldots < a_p$ holds.*

Find the maximum possible value of p and all possible a_1, a_2, ..., a_p sequences.

Solution. There are 32 white squares on a chessboard, so $a_1 + a_2 + \ldots + + a_p = 32$. Since $a_1 \geq 1$, $a_2 \geq 2$, ..., $a_p \geq p$,

$$1 + 2 + \ldots + p \leq a_1 + a_2 + \ldots + a_p = 32;$$

hence the maximal possible value of p is 7. Now, we determine all possible ways of splitting 32 as the sum of 7 positive integers.

$a_3 \geq 3$, but if it was greater, $a_3 + a_4 + \ldots + a_7 \geq 30$ would hold what is impossible. Therefore $a_1 = 1$, $a_2 = 2$, $a_3 = 3$, and $a_1 + a_2 + a_3 = 6$, $a_4 + a_5 + a_6 + a_7 = = 26$.

If $a_5 = 5$, then $a_4 = 4$, and so $a_6 + a_7 = 17$, hence the possible cases are: $6 + 11$, $7 + 10$ and $8 + 9$. We have:

α) $1+2+3+4+5+6+11$,   β) $1+2+3+4+5+7+10$,   γ) $1+2+3+4+5+8+9$.

If $a_5 = 6$, then a_4 equals 4 or 5, in the first case $a_6 + a_7 = 7 + 9$, in the second $a_6 + a_7 = 7 + 8$, so we got:

δ) $1+2+3+4+6+7+9$,   ε) $1+2+3+5+6+7+8$.

$a_5 > 6$ would imply $a_5 + a_6 + a_7 \geq 24$ which is impossible. So we listed the arithmetically possible cases. But α) cannot be realized because for 11 white squares you need a 1 by 22 or a 2 by 11 rectangle. The remaining four cases are realized on *Figure 1974/4.1.*

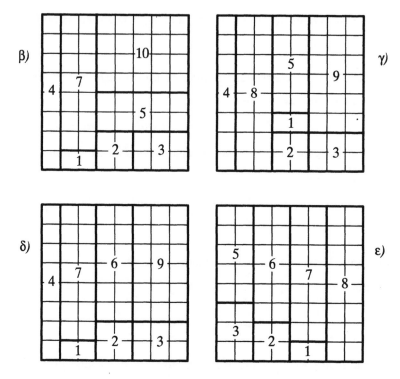

Figure 74/4.1

1974/5. *Determine all possible values of*

$$S = \frac{a}{d+a+b} + \frac{b}{a+b+c} + \frac{c}{b+c+d} + \frac{d}{c+d+a}$$

for arbitrary positive reals a, b, c, d.

Solution. We decrease the sum S if we increase the denominators of the fractions to $a+b+c+d$:

$$S > \frac{a}{a+b+c+d} + \frac{b}{a+b+c+d} + \frac{c}{a+b+c+d} + \frac{d}{a+b+c+d} = 1.$$

We increase S if we decrease the denominator of the first two fractions to $a+b$, and the other two to $c+d$:

$$S < \frac{a}{a+b} + \frac{b}{a+b} + \frac{c}{c+d} + \frac{d}{c+d} = 2,$$

hence

$$1 < S < 2.$$

We show that S attains every value in $]1, 2[$ so this is its range. First, let $a = b = x$ and $c = d = 1$. Set the function

$$S = f(x) = \frac{2x}{2x+1} + \frac{2}{x+2}.$$

$$\lim_{x\to 0+} f(x)=1 \quad \text{and} \quad f(1)=\frac{4}{3},$$

($x\to 0+$ refers to the limit from the right). As f is continuous in $[0,1]$, it attains every λ where $1<\lambda\le\frac{4}{3}$.

Moreover, let $a=c=x$, $b=d=1$; then for

$$S=g(x)=\frac{2x}{x+2}+\frac{2}{2x+1}$$

$$g(1)=\frac{4}{3} \quad \text{and} \quad \lim_{x\to 0+} g(x)=2,$$

$g(x)$ is continuous on the $[0,1]$ interval hence it attains every λ, where $\frac{4}{3}\le\lambda<2$.

The range of the sum S is the $]1,2[$ interval.

Remarks. 1. We used several times the well known theorem of Bolzano, asserting that a continuous function attains every value between its values at the endpoints. Although, we used the argument for semi closed intervals, but we can close them arbitrarily close to 0 or 1, since because of the existence of the limit there is a place in $[0,1]$ where, for example, it attains a value arbitrarily close to 1.

The arguments using continuity are not necessary. For every $1<\lambda\le\frac{4}{3}$ we can find an x, for which $f(x)=\lambda$. Indeed, consider the equation

$$f(x)=\frac{2x}{2x+1}+\frac{2}{x+2}=\lambda.$$

Hence

$$2(\lambda-1)x^2+(5\lambda-8)x+2(\lambda-1)=0,$$

and

$$x=\frac{8-5\lambda+\sqrt{3(\lambda-4)(3\lambda-4)}}{4(\lambda-1)}$$

is a positive solution, because by the restrictions to λ, the inequalities $\lambda-1>0$, $8-5\lambda>0$ hold and the discriminant is nonnegative, too. So there is a positive x such that $f(x)=\lambda$.

Similarly, $g(x)=\lambda$ has a positive solution for $\frac{4}{3}\le\lambda<2$.

2. Instead of f and g we can use the single function $t(x)$: let $a=1$, $b=x$, $c=1-x$, $d=(1-x)x$ ($0<x<1$). Then

$$S=t(x)=\frac{1}{-x^2+2x+1}+\frac{x}{2}+\frac{1-x}{-x^2+x+1}+\frac{(1-x)x}{-x^2+2}.$$

Since

$$\lim_{x\to 0+} t(x)=2, \qquad \lim_{x\to 1} t(x)=1,$$

and t is continuous on $[0,1]$, it attains every λ, where $1<\lambda<2$.

1974/6. *Let $P(x)$ be a non constant polynomial with integer coefficients. Let n be the number of distinct integers k, where $(P(k))^2 = 1$. Prove that*

(1) $$n(P) - \deg(P) \le 2,$$

where $\deg P$ denotes the degree of $P(x)$.

First solution. We start with a widely used observation: if $P(x)$ is a polynomial with integer coefficients, and b, c are distinct integers then $b - c$ divides $P(b) - P(c)$.

Indeed, let

$$P(x) = a_m x^m + a_{m-1} x^{m-1} + \ldots + a_1 x + a_0,$$

then

$$P(b) - P(c) = a_m(b^m - c^m) + a_{m-1}(b^{m-1} - c^{m-1}) + \ldots + a_1(b - c).$$

As $b - c$ divides the right hand side, it also divides the left hand side.

Let m denote the degree of $P(x)$: $m = \deg(P)$; we have to show that

(2) $$n(P) \le m + 2.$$

Assume to the contrary that $n(P) > m + 2$, thus $n(P) \ge m + 3$. In this case there are $m + 3$ distinct integers where the value of $P(x)$ equals $+1$ or -1; let us call them: $k_1 < k_2 < \ldots < k_{m+3}$.

$|P(k_i) - P(k_1)|$ equals 0 or 2, on the other hand $k_2 - k_1 \ge 1$, $k_3 - k_1 \ge 2$, $k_4 - k_1 \ge 3$; but we saw that $k_i - k_1$ divides $P(k_i) - P(k_1)$, thus for $i \ge 4$, $P(k_i) - P(k_1)$ has to be equal to 0. Then the $m + 1$ values

$$P(k_1), \ P(k_4), \ P(k_5), \ \ldots, \ P(k_{m+3})$$

equal which is impossible as a polynomial of degree m can attain the same value at most at m places. Since $n(P) > m + 2$ leads to contradiction so (2).

Second solution. $P(x)$ attains $+1$ or -1 at the integer x if and only if the polynomials $P_1(x) = P(x) - 1$, or $P_2(x) = P(x) + 1$ have an integer root. We show that one of P_1 and P_2 has at most two integer roots.

Assume to the contrary that both P_1 and P_2 have three integer roots. They cannot share a root x_0 in common, because

$$P(x_0) - 1 = P(x_0) + 1$$

is impossible.

Let k be the smallest integer root of P_1 and P_2 e.g. the root of P_1, $P_1(k) = 0$. There is a polynomial $Q(x)$ with integer coefficients such that

(3) $$P_1(x) = (x - k)Q(x).$$

Let a, b and c denote the integer roots of $P_2(x)$, $P_2(a) = P_2(b) = P_2(c) = 0$. The values $Q(a)$, $Q(b)$, $Q(c)$ cannot equal 0, for example, in case $Q(a) = 0$ the equation (3) implies $P_1(a) = 0$, but P_1 and P_2 have no root in common.

As

$$P_1(x) = P_2(x) \perp 2 = (x \perp k)Q(x),$$

substituting a, b, c respectively, we get

$$(a \perp k)Q(a) = (b \perp k)Q(b) = (c \perp k)Q(c) = \perp 2,$$

and so the distinct positive integers $a \perp k$, $b \perp k$, $c \perp k$ divide $\perp 2$, a contradiction, so P_1 and P_2 cannot have both 3-3 roots.

The number of roots of P_1 and P_2 is at most the degree of $P(x)$, that is $\deg(P)$. Thus $P(x)$ can attain $+1$ and $\perp 1$ at at most $\deg(P) + 2$ places, so

$$n(P) \leq \deg(P) + 2,$$

therefore (1) holds.

Remark. Let d denote the difference $n(P) \perp \deg(P)$. Our problem claimed that $d \leq 2$. The truth is that for most polynomials $d \leq 0$. Using the arguments of the first proof this is easy to prove for polynomials of degree at most 5.

Indeed, let us assume that $\deg(P) = m \geq 5$ and $P(x)$ attains $+1$ or $\perp 1$ at the places $k_1 < k_2 < \ldots < k_{m+1}$. Then, for $i > 4$, $k_i \perp k_1$ divides $P(k_i) \perp P(k_1)$ and so $P(k_1) = P(k_4) = \ldots = P(k_{m+1})$; on the other hand, for $i = 1, 2, \ldots, m \perp 2$, $k_{m+1} \perp k_i \geq 3$ and it divides $P(k_{m+1}) \perp P(k_i)$. This implies $P(k_{m+1}) = P(k_2) = = P(k_3)$ and so $P(x)$ takes the same value at $m + 1$ different points, so in case $m \geq 5$, $n(P) \leq m$, thus $d \leq 0$.

We can achieve equality if we consider

$$P(x) = (x \perp 1)(x \perp 2) \ldots (x \perp m) + 1$$

where $(m \geq 5)$.

It can be shown that the only polynomials, for which $n(P) \perp \deg(P) > 0$ are the following ones (here, a denotes an integer):

$$P(x) = \pm \left((x \perp a)^2 + (x \perp a) \perp 1 \right) \qquad d = 2,$$
$$P(x) = \pm \left((x \perp a)(x \perp a \perp 1)(x \perp a \perp 3) + 1 \right) \qquad d = 1,$$
$$P(x) = \pm \left((x \perp a)(x \perp a \perp 2)(x \perp a \perp 3) + 1 \right) \qquad d = 1,$$
$$P(x) = \pm \left(2(x \perp a)^2 \perp 1 \right) \qquad d = 1,$$
$$P(x) = \pm 2(x \perp a) + 1 \qquad d = 1,$$
$$P(x) = \pm (x + a) \qquad d = 1.$$

1975.

1975/1. *Let* $x_1 \geq x_2 \geq \ldots \geq x_n$ *and* $y_1 \geq y_2 \geq \ldots \geq y_n$ *be real numbers. Prove that if* $\{z_i\}$ *is any permutation of the* $\{y_i\}$*, then:*

(1)
$$\sum_{i=1}^{n} (x_i \perp y_i)^2 \leq \sum_{i=1}^{n} (x_i \perp z_i)^2.$$

First solution. Squaring the binomials on both sides and omitting the squares of the x_i-s and y_i-s we get the following inequality equivalent to (1):

$$(2) \qquad \sum_{i=1}^{n} x_i z_i \le \sum_{i=1}^{n} x_i y_i.$$

And this is what we shall prove. In (2) equality holds if the order of the z_i-s agrees with the order of the y_i-s. If this is not the case, let us assume that they first differ at the k-th place, i.e. $y_1 = z_1$, $y_2 = z_2$, ..., $y_{k-1} = z_{k-1}$, but $y_k \ne z_k$. Let $z_k = y_r$ and $y_k = z_s$, where r and s are greater than k, $r > k$ and $s > k$. Using this notation

$$x_k z_k + x_s z_s = x_k y_r + x_s y_k.$$

Now, change the left hand side by substituting $x_k z_k + x_s z_s = x_k y_r + x_s y_k$ by $x_k y_k + x_s y_r$. Doing this, we increased (did not decrease) the sum on the left hand side, because

$$(x_k y_k + x_s y_r) - (x_k y_r + x_s y_k) = (x_k - x_s)(y_k - y_r) \ge 0.$$

With this change we achieved that the first k summands of the left and right hand sides agree. Iterating the procedure, we increase (do not decrease) the left hand side and we arrive to the right hand side, thus verify (2).

Second solution. We shall prove (2), again. Let us introduce the following notations: $A_k = y_1 + y_2 + \ldots + y_k$, $B_k = z_1 + z_2 + \ldots + z_k$, $(k = 1, 2, \ldots, n)$; and by convention let $A_0 = B_0 = 0$. Using these notations we have:

$$y_k = A_k - A_{k-1} \quad \text{and} \quad z_k = B_k - B_{k-1}.$$

Hence (2) becomes:

$$\sum_{k=1}^{n} x_k y_k = \sum_{k=1}^{n} x_k (A_k - A_{k-1}) = \sum_{k=1}^{n} x_k A_k - \sum_{k=0}^{n-1} x_{k+1} A_k =$$

$$= x_n A_n + \sum_{k=1}^{n-1} (x_k - x_{k+1}) A_k.$$

Similarly, we have

$$\sum_{k=1}^{n} x_k z_k = x_n B_n + \sum_{k=1}^{n-1} (x_k - x_{k+1}) B_k.$$

Since A_n and B_n are the sum of the same numbers, $A_n = B_n$. Moreover, $A_k \ge B_k$ $(k = 1, 2, \ldots, n)$ because A_k is the sum of the k largest numbers from these numbers. Thus (2) can be written as

$$\sum_{k=1}^{n-1} (x_k - x_{k+1}) B_k \le \sum_{k=1}^{n-1} (x_k - x_{k+1}) A_k,$$

$$\sum_{k=1}^{n-1}(x_k \perp x_{k+1})(A_k \perp B_k) \geq 0.$$

This is obvious as both factors of the summands of the sum are positive.

Remark. The equation used in the proof says that if $x_1, x_2, \ldots, x_n$ and $z_1, z_2, \ldots, z_n$ are real n-tuples, then the sum

$$S = \sum_{i=1}^{n} x_i z_i$$

is maximal if the tuples are ordered on the same way.

With our methods it is easy to show that the sum is minimal, if the ordering of the z_i-s is the reverse of the ordering of the x_i-s.

From these two theorems several well known inequalities can be derived, as for example the Chebyshev, the Cauchy and the A.M.–G.M. inequalities. (See the remark following our solution of problem 1978/5. See also [35].)

1975/2. *Let $a_1 < a_2 < a_3 < \ldots$ be an infinite sequence of positive integers.*

Prove that there are infinitely many elements of this sequence that can be written in the form

$$a_m = x a_p + y a_q,$$

with x, y positive integers and $p \neq q$.

First solution. Let us assume to the contrary that there are only finitely many elements of the sequence of this form and let a_s be the last one of them. Omit the first s element of the sequence (up to a_s) and denote the remaining elements by

$$b_1, \ b_2, \ b_3, \ \ldots.$$

(If no element is of that form, $b_1 = a_1$). Now, starting at b_2, divide $b_1 + 1$ many elements from the series by b_1. There will be two of them, b_k and b_r ($b_r > b_k$), where the remainders agree, so their difference is divisible by b_1. Hence

$$b_r \perp b_k = x \cdot b_1,$$

where x is a positive integer, so $b_r = x \cdot b_1 + b_k$. Choosing $y = 1$ we get

$$b_r = x \cdot b_1 + y b_k.$$

Thus b_r is of the required form, contradicting our assumption.

Second solution. We shall utilize the following observation (see the Remark): If a, b, c are positive integers such that $c > ab$, where a and b are coprime, then there are positive x and y which satisfy the equation $ax + by = c$.

Let d_i denote the greatest common divisor of a_1 and a_i ($i = 2, 3, \ldots$). Since a_1 has only finitely many divisors, there is a number in the sequence $d_2, d_3, \ldots$

that occurs infinitely many times, so in the a_i sequence there are infinitely many elements of the form

$$a_1 = b_1 d, \quad b_2 d, \quad b_3 d, \quad \ldots$$

The sequence b_i is strictly increasing, so it contains infinitely many elements greater than $b_1 b_2$. The greatest common divisor of $a_1 = b_1 d$ and $b_2 d$ is d, implying that b_1 and b_2 are coprime. So, by our introductory observation, the equation

$$b_1 x + b_2 y = b_k$$

has a positive integer solution along with the equation

$$b_1 dx + b_2 dy = b_k d,$$

where $b_1 d$, $b_2 d$, $b_k d$ are in the sequence a_i. There are infinitely many choices for b_k, therefore the statement is proved.

Remark. We prove the observation in the second solution: the integers a, $2a$, $3a$, $\ldots$, ba give distinct residues mod b. Indeed, in case ia and ja $(1 \le i < < j \le b)$ give the same residue, then, b divides $(j \perp i)a$ that is impossible since $j \perp i < b$ and a and b are coprime. Hence the numbers a, $2a$, $\ldots$, ba give all possible residues mod b, they form a so called complete residue system.

So, one of them — e.g. xa —, gives the same residue as c, and so b divides $c \perp ax$, hence there is an integer y, such that

$$c \perp ax = by.$$

As $x \le b$, we have $ax \le ab < c$ and $c \perp ax = by > 0$, so $y > 0$ and the equation $ax + by = c$ has a positive solution x, y.

1975/3. *Given any triangle ABC, construct external triangles ABR, BCP, CAQ on the sides, so that*

$$\angle PBC = \angle CAQ = 45°$$
$$\angle BCP = \angle QCA = 30°$$
$$\angle ABR = \angle BAR = 15°.$$

Prove that $\angle QRP = 90°$ and $QR = RP$.

First solution. Choose the notations such that the triangles ABC, BPC, CQA ARB are labelled counterclockwise (see *Figure 1975/3.1*).

$$\frac{AC}{AQ} = \frac{BC}{BP} = \lambda,$$

because the triangles CQA and CPB are similar. Let τ_1 denote the rotation-stretching with ratio λ, degree $\perp 45°$ and centre A, τ_2 the rotation-stretching with ratio $\dfrac{1}{\lambda}$, angle $\perp 45°$ and centre B. τ_1 maps Q to C, τ_2 maps C to P, so the composition of them, $\tau_1 \tau_2$ maps Q to P. $\tau_1 \tau_2$ is a rotation-stretching with ratio $\lambda \cdot \dfrac{1}{\lambda} = 1$, angle $2 \cdot 45° = 90°$, so it is a congruence, thus a rotation by $90°$.

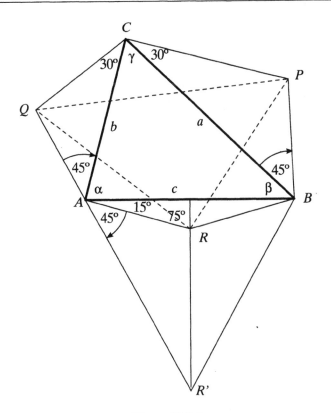

Figure 75/3.1

The transformation τ_1 maps R to a point R', such that the triangles AQC and ARR' are similar. Thus $\angle ARR' = 105°$, hence RR' is orthogonal to AB. Since ARB is an isosceles triangle, RR' is the axis of symmetry of the triangle ABR', therefore ABR' is an isosceles triangle (it is equilateral as $\angle BAR' = 60°$). This implies that ARR' and BRR' are congruent triangles, hence τ_2 maps R' to R. Thus R is the fixpoint of $\tau_1\tau_2$ and so its centre. Hence $\tau_1\tau_2$ is a rotation by $90°$ with centre R mapping Q to P. Thus QRP is an isosceles right triangle and this is what we wanted to prove.

Second solution. Trigonometry provides another solution.

Applying the Law of Sines to the triangles ACQ and PBC we get

$$\frac{AQ}{b} = \frac{BP}{a} = \frac{\sin 30°}{\sin 105°} = \frac{1}{2\sin 75°}; \quad \frac{CQ}{b} = \frac{CP}{a} = \frac{\sin 45°}{\sin 105°} = \frac{1}{\sqrt{2}\sin 75°},$$

hence we obtain

(1) $\quad AQ = \dfrac{b}{2\sin 75°}, \quad BP = \dfrac{a}{2\sin 75°}, \quad CQ = \dfrac{b}{\sqrt{2}\sin 75°}, \quad CP = \dfrac{a}{\sqrt{2}\sin 75°}.$

In the isosceles triangle ARB we have

(2) $$AR = BR = \frac{c}{2\sin 75°}.$$

Applying the Law of Cosine to QAR, PBR and PQC using (1) and (2) we get:

$$QR^2 = AQ^2 + AR^2 - 2AQ \cdot AR \cos(\alpha+60°) =$$

$$= \frac{1}{4\sin^2 75°}\left(b^2 + c^2 - 2bc\cos(\alpha+60)\right).$$

Similarly,

$$PR^2 = \frac{1}{4\sin^2 75°}\left(a^2 + c^2 - 2ac\cos(\beta+60°)\right),$$

$$PQ^2 = \frac{1}{2\sin^2 75°}\left(a^2 + b^2 - 2ab\cos(\gamma+60°).\right)$$

We show that the expressions in the latter three parentheses represent the same value. We will express them with the area of the triangle in the same way, e.g.:

$$S = a^2 + b^2 - 2ab\cos(\gamma+60°) = a^2 + b^2 - 2ab\left(\frac{1}{2}\cos\gamma - \frac{\sqrt{3}\sin\gamma}{2}\right) =$$

$$(3) \qquad = a^2 + b^2 - \frac{a^2 + b^2 - c^2}{2} + \sqrt{3}ab\sin\gamma = \frac{a^2 + b^2 + c^2}{2} + 2\sqrt{3}t.$$

Clearly, we get the same formula starting with any of the three expressions, so

$$QR = PR \qquad \text{and} \qquad PQ^2 = QR^2 + PR^2,$$

therefore PQR is an isosceles right triangle. (We also remark that the formula used in the last step of the solution also follows from the so called isogonal point theorem, see [4]).

Third solution. The following generalization goes deeper into the background of the problem. Two 4-tuples of points are called similar if there is a similarity transformation mapping one to the other. If you pick 3-3 corresponding vertices in two similar 4-tuples, the triangles spanned by them are similar. Two similar 4-tuples have the same orientation if the corresponding triangles have the same orientation. Theorem:

If $XYZU$, $X'Y'Z'U'$, $X''Y''Z''U''$ are similar 4-tuples and $Z' = Y$, $Y'' = X'$, $Z'' = X$, moreover $U' \neq U$, $U'' \neq U$, then the triangles $XX'X''$, $YY'Y''$, $ZZ'Z''$ and $UU'U''$ are pairwise similar (see *Figure 1975/3.2*).

Visually, if there is an arbitrary triangle on the plane (XYX' on the figure) and we insert three similar 4-tuples on the following way: the XY edge of the first 4-tuple is the same as the XY edge of the triangle; the $Z'X'$ edge of the second 4-tuple is the same as the $YX' = Z'Y''$ edge of the triangle; the $Y''Z''$ edge of the third 4-tuple is the same as the $X'X = Y''Z''$ edge of the triangle, then the four triangles mentioned in the theorem are pairwise similar.

The similarity of the first three triangles is trivial. Indeed, XYZ is identical to $Z''Z'Z$ and $X'Y'Z'$ is identical to $Y''Y'Y$; it only remains to show that $UU'U''$ is similar to the first three triangles.

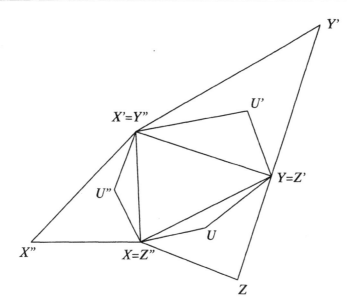

Figure 75/3.2

For the proof, we insert our figure to the complex plain with U as the 0; we denote the complex numbers belonging to the points with the corresponding lower case letters. We do not distinguish the points from the corresponding numbers .

Every similarity keeping the orientation mapping the point p to p^* is given by a linear function. Let

$$(4) \qquad p^* = k_1 p + t_1,$$

be the transformation mapping $XYZU$ to $X'Y'Z'U'$ and

$$(5) \qquad p^* = k_2 p + t_2,$$

the one mapping $XYZU$ to $X''Y''Z'U''$, where k_1, t_1, k_2, t_2 are the appropriate complex numbers.

As $u = 0$ and (4) maps u to u', and (5) to u'', we have $u' = k_1 \cdot 0 + t_1 = t_1$ and $u'' = k_2 \cdot 0 + t_2 = t_2$, so (4) and (5) become

$$(6) \qquad p^* = k_1 p + u', \qquad \text{and} \qquad p^* = k_2 p + u''.$$

Now, apply (6) to the $x \to x'$, $z \to z' = y$, and $y \to y'' = x'$, $z \to z'' = x$ correspondences:

$$(7) \qquad x' = k_1 x + u' \qquad\qquad (9) \qquad x' = k_2 y + u''$$

$$(8) \qquad y = k_1 z + u' \qquad\qquad (10) \qquad x = k_2 z + u''$$

As $x \neq 0$, $y \neq 0$, (7) and (9) gives:

$$k_1 = \frac{x' \perp u'}{x}, \qquad k_2 = \frac{x' \perp u''}{y}.$$

Substituting to (8) and (10) we get:

$$y = \frac{x'z \perp u'z + u'x}{x}, \qquad x = \frac{x'z \perp u''z + u''y}{y}.$$

Combining the two formulas:

$$xy = x'z \perp u'z + u'x = x'z \perp u''z + u''y; \quad u'(x \perp z) = u''(y \perp z).$$

Now, set:

$$k_3 = \frac{x \perp z}{u''} = \frac{y \perp z}{u'} \qquad (u' \neq 0, u'' \neq 0.)$$

Applying the similarity transformation $p^* = k_3 p + z$ to u, u', and u'', respectively, we obtain

$$k_3 u + z = z,$$
$$k_3 u' + z = y,$$
$$k_3 u'' + z = x.$$

That is $U''U'U$ is similar to XYZ and this is what we wanted to prove.

Now, if we consider the $XYZU$ 4-tuple where XYZ is an isosceles right triangle, (the right angle is at Z) and inside take U such that XUY is an isosceles triangle with $\angle ARU = 150°$, then we get the configuration of the problem.

Remarks. 1. The method of the first proof gives us the following generalisation:

If we construct the BPC, CQA, ARB triangles outside to the arbitrary ABC triangle such that

$$\angle PBC = \angle CAQ = \varphi,$$
$$\angle BCP = \angle QCA = \delta,$$
$$\angle ABR = \angle BAR = |90° \perp (\varphi + \delta)|$$

and AB separates R and C if $\varphi + \delta < 90°$; AB does not separate them if $\varphi + \delta > 90°$; finally, R is the midpoint of AB if $\varphi + \delta = 90°$, then $\angle QRP = 2\varphi$ and $QR = RP$.

In our case $\varphi = 45°$, $\delta = 30°$. This formulation of the problem has several interesting special cases. For example, if $\varphi = \delta = 45°$, then we construct isosceles right triangles to the AC and BC sides of the ABC triangle; in this case P, Q and the midpoint of AB form a isosceles right triangle.

2. A historically interesting special case of the generalisation above: if $\varphi = \delta = 30°$. The vertices of the isosceles triangles are the centres of the equilateral triangles constructed above the sides. By the result, they form an isosceles triangle with an angle of degree $60°$, and this is an equilateral triangle. This problem is due to Napoleon.

An other generalisation is the Napoleon–Barlotti theorem: if we construct regular n-gons on the sides of an n-gon, the centres of these n-gons form a regular n-gon if and only if the original n-gon was a regular affine n-gon. (An n-gon is affine regular if it is the projection of a regular n-gon. eg. every triangle is affine regular; the affine regular quadrilaterals are the parallelograms.)

1975/4. *Let A denote the sum of the decimal digits of 4444^{4444}, and B be the sum of the decimal digits of A. Find the sum of the decimal digits of B.*

Solution. Let N denote the number 4444^{4444}, and C the sum of the decimal digits of B. As the mod 9 residue of a number is the same as the residue of the sum of its digits, N gives the same residue mod 9 as A, B and C. 4444 gives residue 7, so $N = 9k + 7$, for some positive integer k. The powers of $9k + 7$ are of the form: $(9k + 7)^2 = 9k_1 + 4$, $(9k + 7)^3 = 9k_2 + 1$, $(9k + 7)^4 = 9k_3 + 7$, so the mod 9 residues are periodic, 7, 4, 1, 7, $\ldots$ If the exponent is $3m + 1$, $3m + 2$, $3m$ the mod 9 residues are 7, 4, 1, respectively. Since 4444 is of the form $3m + 1$, N gives 7 mod 9 and 7 is the residue of A, B and C.

As $N < 10\,000^{4444} = 10^{4 \cdot 4444} = 10^{17\,776}$, N has at most $17\,776$ decimal digits. Thus, $A \le 9 \cdot 17\,776 = 159\,984$. The first digit of A is at most 1, hence $B \le \le 1 + 5 \cdot 9 = 46$. Among the first 46 number the sum of the digits is the largest in case of 39. Thus $C \le 12$, but C gives residue 7 mod 9, and so $C = 7$.

Remarks. 1. N has $16\,211$ digits, $A = 72\,601$, $B = 16$ finally, $C = 7$.

2. We can get the mod 9 residue of N using congruences:
$$N = 4444^{4444} \equiv 7^{4444} \equiv (\bot 2)^{3 \cdot 1481 + 1} \equiv 7 \cdot (\bot 8)^{1481} \equiv 7 \cdot 1^{1481} = 7.$$

1975/5. *Can you find 1975 points on the circumference of a unit circle such that the distance between each pair is rational?*

First solution. We show that we can choose n points on the unit circle such that the distance of any two of the points is rational $(n > 1)$.

We shall utilize the existence of an arbitrary small angle with rational tangent. Indeed, let in the ABC right triangle 1 be the length of the BC leg and N (positive integer) the length of the AC hypotenuse, $\angle BAC = \varphi$. Then $\tan \varphi = \dfrac{1}{N}$ is rational; increasing N, we can make $\tan \varphi$ and φ arbitrarily small.

Now, choose the angle α such that $\alpha < \dfrac{\pi}{n \bot 1}$ with $\tan \dfrac{\alpha}{2}$ rational. In this case $\sin \alpha$ and $\cos \alpha$ are rational, too, as

$$\sin \alpha = \frac{2 \tan \frac{\alpha}{2}}{1 + \tan^2 \frac{\alpha}{2}} \qquad \text{and} \qquad \cos \alpha = \frac{1 \bot \tan^2 \frac{\alpha}{2}}{1 + \tan^2 \frac{\alpha}{2}},$$

moreover, $\sin n\alpha$ and $\cos n\alpha$ are rational, as well. This can be proved by induction using that

$$\sin(n + 1)\alpha = \sin n\alpha \cos \alpha + \cos n\alpha \sin \alpha,$$
$$\cos(n + 1)\alpha = \cos n\alpha \cos \alpha \bot \sin n\alpha \sin \alpha.$$

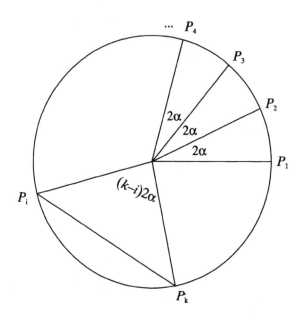

Figure 75/5.1

For $n = 1$ the statement is true, and if it is true for n, the above formulae prove it for $n + 1$.

Now, consider the points P_1, P_2, P_3, ..., P_n on the unit circle such that the angle between two neighbouring points is 2α (see *Figure 1975/5.1*). These points are distinct, because $(n \perp 1) \cdot 2\alpha < 2\pi$. The length of the arc between P_i and P_k $(k > i)$ is $(k \perp i) \cdot 2\alpha$, hence

$$P_i P_k = 2 \sin(k \perp i)\alpha,$$

and as $\sin \alpha$ is rational, the length of $P_i P_k$ is rational, too.

Second solution. Construct right triangles over the segment $P_1 P_n$ as hypotenuse of length 2, with legs:

$$P_1 P_k = \frac{4k}{k^2 + 1}, \quad P_k P_n = \frac{2(k^2 \perp 1)}{k^2 + 1} \quad (k = 2, 3, \ldots, n \perp 1).$$

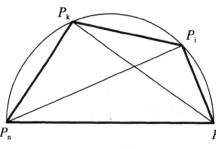

Figure 75/5.2

Clearly, $P_1 P_k^2 + P_k P_n^2 = 4$, so these triangles exist, moreover their P_k vertices are on the unit circle constructed over $P_1 P_k$. The points P_1, P_2, ..., P_n satisfy the conditions of the problem, because $P_1 P_k$ and $P_k P_n$ are rational by definition. Now, applying Ptolemy's theorem ([24]) for $1 < i < k < n$ (see *Figure 1975/5.2*) we get:

$$P_1 P_k \cdot P_n P_i = P_1 P_n \cdot P_i P_k + P_1 P_i \cdot P_n P_k,$$

and

$$P_i P_k = \frac{1}{2}(P_1 P_k \cdot P_n P_i \perp P_1 P_i \cdot P_n P_k).$$

As the right hand side is the product and sum of rational numbers, $P_i P_k$ is rational, too.

Remark. The theorem implies that there are N points in the plane such that no 3 of them are collinear, but the distance of any 2 of them is an integer. Indeed, if the distances in the construction are

$$\frac{p_1}{q_1}, \quad \frac{p_2}{q_2}, \quad \dots, \quad \frac{p_r}{q_r}$$

enlarging the plane by $q_1 q_2 \dots q_r$, the distances become integers.

It is surprising, however, that there cannot be found infinitely many points on the plane such that the distance of any two is an integer. To show this, let $AB = c$, $BC = a$, $CA = b$ in the ABC triangle, where a, b, c are integers. We show that there are only finitely many points D on the plane such that its distance is integer from the vertices A, B, and C.

From the triangle inequality we have:

$$|DB \perp DC| \leq a, \qquad |DA \perp DB| \leq c.$$

The set of points, D where $|DB \perp DC| = k$ (constant) is a hyperbola with foci B and C $(0 < k < a)$, or the perpendicular bisector of BC $(k = 0)$, or the BC line, omitting the segment BC $(k = a)$. Since a is an integer, k has to be 0, 1, 2, ..., a so D is situated on one of the two lines or $a \perp 1$ hyperbolas. Similarly, $|DA \perp DB| \leq c$ implies that D is lies on one of the two lines or $c \perp 1$ hyperbolas. Two hyperbolas or pair of lines intersect in at most four points. Thus there are finitely many positions for D, and it is impossible to find infinitely many points with integer distances under the conditions of the problem.

1975/6. *Find all polynomials $P(x, y)$ in two variables such that:*

I. *for every real numbers t, x, y, $P(tx, ty) = t^n P(x, y)$, where n is a positive integer, i.e. P is a homogeneous polynomial of degree n.*

II. *For every real a, b, c, $P(a + b, c) + P(b + c, a) + P(c + a, b) = 0$.*

III. *$P(1, 0) = 1$.*

Solution The general form of a homogeneous polynomial of degree n is

(1) $P(x, y) = a_n x^n + a_{n-1} x^{n-1} y + a_{n-2} x^{n-2} y^2 + \dots + a_1 x y^{n-1} + a_0 y^n$.

(1) (or property I) implies that $P(0, 0) = 0$. If the degree of $P(x, y)$ is 1, then P is of the form $a_n x + a_0 y$, and from II and III we have $x = 2y$. We show that in

case $n > 1$, for every real x, the identity $P(\perp x, x) = 0$ holds. Applying property II by substituting $a = b = x$ and $c = \perp 2x$ gives:

$$P(2x, \perp 2x) + P(\perp x, x) + P(\perp x, x) = 0,$$

$$((\perp 2)^n + 2)P(\perp x, x) = 0,$$

$$P(\perp x, x) = 0.$$

This says that if $x + y = 0$, than

(2) $$P(x, y) = 0.$$

Now, assume that $x + y \neq 0$. Applying II with $c = 1 \perp a \perp b$ we obtain:

(3) $$P(a + b, 1 \perp a \perp b) + P(1 \perp a, a) + P(1 \perp b, b) = 0.$$

Substituting $b = 0$ we get:

$$P(a, 1 \perp a) + P(1 \perp a, a) + 1 = 0$$

Hence

$$P(1 \perp a, a) = \perp P(a, 1 \perp a) \perp 1.$$

As (3) is symmetric in a and b we have

$$P(1 \perp b, b) = \perp P(b, 1 \perp b) \perp 1.$$

Substituting to (3) and reordering the expression we get

(4) $$P(a + b, 1 \perp a \perp b) = P(a, 1 \perp a) + P(b, 1 \perp b) + 2.$$

Introducing $Q(z) = P(z, 1 \perp z) + 2$, (4) gives

$$Q(a + b) = Q(a) + Q(b),$$

where $Q(z)$ is a polynomial of one variable. From III. we get $Q(1) = P(1, 0) + 2 = 3$, and with induction, $Q(n) = 3n$, because $Q(n + 1) = Q(n) + Q(1) = 3n + 3 = 3(n + 1)$. The values of $Q(z)$ and $3z$ agree at infinitely many places, hence

$$Q(z) = 3z, \quad \text{and so} \quad P(z, 1 \perp z) = 3z \perp 2.$$

This gives

$$P(x, y) = (x + y)^n P\left(\frac{x}{x+y}, \frac{y}{x+y}\right) = (x + y)^n P\left(\frac{x}{x+y}, 1 \perp \frac{x}{x+y}\right) =$$

$$= (x + y)^n \left(\frac{3x}{x+y} \perp 2\right) = (x + y)^{n-1}(x \perp 2y),$$

and this satisfies (2) if $x + y = 0$, as well.

The polynomial obviously satisfies I and III. Condition II holds, as

$$(a + b + c)^{n-1}(a + b \perp 2c) + (a + b + c)^{n-1}(b + c \perp 2a) + (a + b + c)^{n-1}(c + a \perp 2b) = 0.$$

Thus the only polynomial satisfying the conditions of the problem:

$$P(x, y) = (x + y)^{n-1}(x \perp 2y).$$

Remark. Using additional properties of the polynomials (e.g. continuity) there are different ways to solve the problem. We sketch one of these possibilities:

If $n > 1$, by (2) $P(x, \perp x) = 0$. Fix x and consider $P(x, y) = P_x(y)$ that is a one variable polynomial of y. As $P(x, \perp x) = 0$, $x + y$ is a factor of $P_x(y)$:

$$P(x, y) = (x + y)Q_1(x, y),$$

where $Q_1(x, y)$ is a homogeneous polynomial of two variables.

$$0 = P(a + b + c) + P(b + c, a) + P(c + a, b) =$$

$$= (a + b + c)\big(Q_1(a + b, c) + Q_1(b + c, a) + Q_1(c + a, b)\big),$$

so in case $a + b + c \neq 0$, $Q_1(x, y)$ satisfies II.

In case $x + y \neq 0$, $Q_1(x, y)$ satisfies (1), because

$$Q_1(tx, ty) = \frac{P(tx, ty)}{t(x + y)} = t^{n-1}Q_1(x, y),$$

that is $Q_1(x, y)$ satisfies the conditions I–III. Now, using the continuity of $P(x, y)$ this holds if $a + b + c = 0$, or $x + y = 0$. Applying the same procedure to $Q_1(x, y)$, we can factor out $x + y$ from $Q_1(x, y)$ until we get the polynomial $Q_i(x, y)$ of degree 1, that has to be of the form $x \perp 2y$, thus

$$P(x, y) = (x + y)^{n-1}(x \perp 2y).$$

7. A Glossary of Theorems

[1] *The paralelogram theorem and an application.* The sum of the squares of a paralelogram's diagonals is equal to that of the sides. Denote, for the proof, the vectors spanning the paralelogram by **a** and **b**; hence its diagonals are $\mathbf{a}+\mathbf{b}$ and $\mathbf{a}\perp\mathbf{b}$, respectively. The claim is now straightforward as

$$2\mathbf{a}^2 + 2\mathbf{b}^2 = (\mathbf{a}+\mathbf{b})^2 + (\mathbf{a}\perp\mathbf{b})^2.$$

Let the lengths of the sides of a triangle be a, b and c and that of the median $CC' = s_c$ (*Figure 1.1.*). Reflecting the triangle in C' yields a paralelogram of sides a, b, a, b; its diagonals are c and $2s_c$. By the previous result

$$2a^2 + 2b^2 = 4s_c^2 + c^2,$$

and hence

$$s_c^2 = \frac{2a^2 + 2b^2 \perp c^2}{4}.$$

[2] *Sides and cotangents in a triangle.*

$$a^2 + b^2 + c^2 = 4t(\cot\alpha + \cot + \cot\gamma).$$

Write down the cosine rule for the sides and sum the equalities:

$$a^2 + b^2 + c^2 = 2\left(a^2 + b^2 + c^2\right) \perp 2bc\cos\alpha \perp 2ca\cos\beta \perp 2ab\cos\gamma,$$

$$a^2 + b^2 + c^2 = 2\left(bc\cos\alpha + ca\cos\beta + ab\cos\gamma\right).$$

From the area formula: $bc = \dfrac{2A}{\sin\alpha}$; substituting the symmetric permutations of this relation into the previous result yields the claim. Indeed

$$a^2 + b^2 + c^2 = 4t\left(\frac{\cos\alpha}{\sin\alpha} + \frac{\cos\beta}{\sin\beta} + \frac{\cos\gamma}{\sin\gamma}\right) = 4t(\cot\alpha + \cot\beta + \cot\gamma).$$

[3] *The cotangent inequality.* If α, β, γ are the angles of a triangle then

$$\cot\alpha + \cot\beta + \cot\gamma \geq \sqrt{3}.$$

There are several ways to prove this inequality; starting with straightforward identities we proceed by simple estimations.

$$\cot\alpha + \cot = \frac{\cos\alpha}{\sin\alpha} + \frac{\cos\beta}{\sin\beta} = \frac{\sin(\alpha+\beta)}{\sin\alpha\sin\beta} = \frac{2\sin\gamma}{\cos(\alpha\perp\beta)\perp\cos(\alpha+\beta)} =$$

$$= \frac{2\sin\gamma}{\cos(\alpha\perp\beta)+\cos\gamma} \geq \frac{2\sin\gamma}{1+\cos\gamma} = \frac{4\sin\frac{\gamma}{2}\cos\frac{\gamma}{2}}{1+2\cos^2\frac{\gamma}{2}\perp 1} = 2\tan\frac{\gamma}{2}.$$

Therefore

$$\cot\alpha + \cot\beta + \cot\gamma \geq \cot\gamma + 2\tan\frac{\gamma}{2} =$$

$$= 2\tan\frac{\gamma}{2} + \frac{\cot^2\frac{\gamma}{2}\perp 1}{2\cot\frac{\gamma}{2}} = \frac{1}{2}\frac{\cot^2\frac{\gamma}{2}+3}{\cot\frac{\gamma}{2}} = \frac{1}{2}\left(\cot\frac{\gamma}{2} + 3\tan\frac{\gamma}{2}\right).$$

The last sum can be estimated from below by the A.M.–G.M. inequality.

$$\cot\alpha+\cot\beta+\cot\gamma\geq\sqrt{\cot\frac{\gamma}{2}\cdot 3\tan\frac{\gamma}{2}}=\sqrt{3}.$$

[4] *Isogonal point* (isogonal = equiangular). The sides of a triangle are subtending equal angles at the isogonal point. This common angle is clearly equal to $120°$ and one can find such a point if the angles of the triangle are all less than $120°$.

The isogonal point is hence incident to the circular arcs through the endpoints of the sides corresponding to $120°$; these arcs do necessarily have a point in common.

The following simple construction is also leading to the isogonal point. Draw equilateral triangles ABC', BCA' and CAB' above the sides of the triangle ABC externally; the line segments AA', BB', CC' are then congruent (*Figure 4.1.*). The segments AA' and BB', for example, are equal because the rotation by $60°$ about C is mapping the triangle ACA' into $B'CB$; this is true if these triangles are degenerated into a segment.

If the isogonal point I does exist then it is incident to each of the segments AA', BB', CC'. In fact, since $AIC\angle=AIB\angle=120°$ rotating the triangle AIB about A by $60°$ one obtains the triangle $AI'C'$. By the same rotation the triangle AII' is equilateral (*Figure 4.2.*). The diagram reveals that the points C, I and C' are collinear, therefore I is incident to the segment CC', indeed. Apart from that $CC'=IA+IB+IC$ and thus $CC'(=AA'=BB')$ is equal to the sum of the distances of the point I from the vertices of the triangle.

Starting with a point I not lying on the segment CC' the previous rotation yields the following inequality:

$$AI+BI+CI=II'+CI+I'C'>CC',$$

since the segments II', CI, $I'C'$ now form a triangle. Differently speaking, the isogonal point – if it exists – is minimizing the sum of the distances of a point from the vertices of a triangle; this minimum is actually equal to the common length of the segments AA', BB', CC'.

When written down in the triangles BCB', CAC', ABA', respectively, the cosine rule yields

$$BB'=CC'=AA'=a^2+b^2\perp 2ab\cos(\gamma+60°)=$$
$$=b^2+c^2\perp 2bc\cos(\alpha+60°)=c^2+a^2\perp 2ac\cos(\beta+60°).$$

This latter equality, by the way, can be obtained in a straightforward manner without the above investigations.

[5] *Bicentric polygons; Poncelet's porisms.* A polygon is called bicentric if it has both an inscribed and a circumscribed circle. Any triangle is bicentric, for example, and also all the regular polygons. There is a relation, for bicentric polygons,

between the inradius r, the circumradius R and the distance d of the respective centres depending also on the number of sides. Here there is a list of the first few of them for 3, 4, 5 and 6 sided polygons.

$n = 3$: $d^2 = R^2 \perp 2Rr$ or: $\dfrac{1}{R+d} + \dfrac{1}{R \perp d} = \dfrac{1}{r}$,   (Euler)

$n = 4$: $\left(R^2 \perp d^2\right)^2 = 2r^2\left(R^2 + d^2\right)$ or: $\dfrac{1}{(R+d)^2} + \dfrac{1}{(R \perp d)^2} = \dfrac{1}{r^2}$,

$n = 5$: $r(R \perp d) = (R+d)iR \perp r \perp d\,(iR \perp r + d + i2R)$,

$n = 6$: $3\left(R^2 \perp d^2\right)^4 = 4r^2\left(R^2 + d^2\right)\left(R^2 \perp d^2\right) + 16R^2d^2r^4$.

The circles of bicentric polygons have the celebrated property that they are, in fact, shared by infinitely many bicentric $n$gons; to put it more precisely denote the incircle and the circumcircle of a bicentric polygon by c and C, respectively. Draw from an arbitrary point A_1 of C a tangent to c and let this line meet C at A_2. The – other – tangent from A_2 to c meets C at A_3, etc. The point A_{n+1} of this process is then incident to A_1, the sequence of segments hence drawn is closing up at the nth step yielding the porism of Poncelet in circles.

In the general form of Poncelet's theorem there are conic sections for circles; accordingly, the general theorem is a gem of projective geometry.

[6] *Power of a point with respect to a circle; radical axis; radical centre.* Given is the circle c of radius R and centre O, for an arbitrary point P in the plane of c the real number

$$h = PO^2 \perp R^2$$

is called the power of the point P with respect to the circle c. This number is positive, zero or negative if the point P is outside of c, lying on c or it is inside c. For external points the value of h is equal to the square of the tangent from P to the circle.

The locus of the points whose power is equal with repect to two non concentric circles is a straight line perpendicular to their axis. This line is called the radical axis of the two circles. It is the line of the common chord if the circles intersect; it is the common – internal – tangent if they touch; in general, it contains those points from where one can draw equal tangents to the circles.

In case of three circles the pairwise drawn radical axes are either parallel or they meet at a common point; in the latter case this point is called the radical centre of the three circles.

[7] *The position vector of the incentre.* Denote the position vectors of the vertices of the triangle ABC by **a**, **b** and **c**, respectively. Since the intersection C_1 of the bisector of the angle C and the opposite side AB is dividing AB in the ratio $b:a$

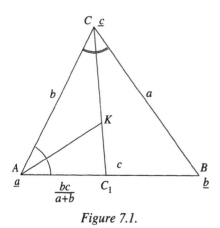

Figure 7.1.

(*Figure 7.1.*), the position vector of C_1 is equal to

$$\frac{a\mathbf{a}+b\mathbf{b}}{a+b}.$$

The bisector of the angle A meets the segment CC_1 at the incenter K dividing CC_1 in the ratio $b:\dfrac{bc}{a+b}=1:\dfrac{c}{a+b}$. K's position vector is hence

$$\mathbf{k}=\frac{\frac{c}{a+b}\cdot\mathbf{c}+\frac{a\,\mathbf{a}+b\,\mathbf{b}}{a+b}}{1+\frac{c}{a+b}}=\frac{a\,\mathbf{a}+b\,\mathbf{b}+c\,\mathbf{c}}{a+b+c}.$$

[8] *The inradius and the circumradius of a triangle.* There are several formulas for these radii in terms of the sides and the angles of the triangle; here there are but a few of them. The inradius is denoted by r, the circumradius is by R, the area of the triangle by A, the sides and the angles by a, b, c, and α, β, γ, respectively and finally, the semiperimeter by s.

$$R=\frac{abc}{4A}=\frac{a}{2\sin\alpha}=\frac{b}{2\sin\beta}=\frac{c}{2\sin\gamma},$$

$$R^2=\frac{2A}{\sin 2\alpha+\sin 2\beta+\sin 2\gamma}$$

$$r=\frac{A}{s}=4R\sin\frac{\alpha}{2}\sin\frac{\beta}{2}\sin\frac{\gamma}{2},$$

$$r^2=A\tan\frac{\alpha}{2}\tan\frac{\beta}{2}\tan\frac{\gamma}{2}.$$

[9] *Poncelet's theorem* see [5].

[10] *Touching circles.* Two circles in the space are touching each other if they have a single point in common and they share the tangent at this point. The axis of a circle is the line perpendicular to its plane at the centre. The axis contains the points whose distance from the points of the circle are all equal.

The plane perpendicular to the common tangent at the point T_{12} of contact of the touching circles c_1 and c_2 is containing the axes of both; if the circles are not coplanar then the axes meet and the distance of their common point O is equal to OT_{12} from the points of both circles; they are hence incident to the sphere of centre O and radius OT_{12}.

Consider now three pairwise touching circles c_1, c_2, c_3 that are not coplanar. Denote the touching points by T_{12}, T_{13}, T_{23}, respectively. Let the sphere containing c_1 and c_2 be S. Since there is a unique sphere passing through a circle c and a point not lying on c, the circle c_1 and the point T_{23} determine the sphere S; this sphere is also passing through the circle c_3. Accordingly, three pairwise touching circles, if not coplanar, are lying on a sphere.

11] *Equilateral tetrahedron* A tetrahedron is called equilateral if its faces are congruent. Being the face of an equilateral tetrahedron a triangle is always acute angled; any such triangle, on the other hand, is the face of some equilateral tetrahedron.

  The opposite edges of an equilateral tetrahedron are pairwise equal; its circumscribed parallelepiped is hence a cuboid. Its specific points, namely the incentre, the circumcentre and its centroid are incident; conversely, if any two of the above points are incident then the tetrahedron is equilateral. Finally, if the areas of the faces of a tetrahedron are equal then the terahedron is equilateral.

12] *The cosine inequality.* As a fundamental inequality in triangles it is the source of several further results; it states that

$$\cos\alpha + \cos\beta + \cos\gamma \leq \frac{3}{2},$$

and equality holds if and only if the triangle is equilateral.

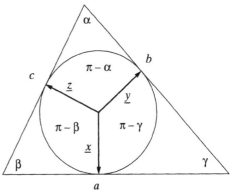

  The following non standard argument is using vectors: set the origin as the incentre and let the inradius be equal to 1. The unit vectors to the points of contact of the incircle are denoted by **x**, **y**, **z**, respectively. Using the notations of *Figure 12.1*

Figure 12.1.

$$0 \leq (x+y+z)^2 = 3 + 2xy + 2yz + 2zx =$$

$$= 3 + 2\left(\cos(a\pi \perp \gamma) + \cos(a\pi \perp \alpha) + \cos(a\pi \perp \beta)\right) = 3 \perp 2\left(\cos\alpha + \cos\beta + \cos\gamma\right),$$

yielding

$$\cos\alpha + \cos\beta + \cos\gamma \leq \frac{3}{2},$$

and equality holds if and only if $x+y+z=0$, that is $1 = x^2 = (\perp y \perp z)^2 = 2 + 2\cos(a\pi \perp \alpha)$, $\cos(a\pi \perp \alpha) = \perp\frac{1}{2}$, $180° \perp \alpha = 120°$. The pairwise angles of the vectors **x**, **y**, **z** are $120°$, the triangle is equilateral.

13] *Ramsey's theorem.* If the edges of an n-point graph are coloured with two colours say blue and red then Ramsey's theorem claims the existence of a monochromatic complete subgraph of certain size. Formally:

  For every pair (b, r) of positive integers there exists a positive integer $R(b, r)$ such that if $n \geq R(b, r)$ and each edge of a complete n-graph is coloured either blue or red then either there is a complete subgraph of b vertices whose edges are all blue or there is a complete subgraph of r vertices whose edges are all red. If, on the other hand, $n < R(b, r)$, then one can colour the edges of a complete

n-graph in such a way that there are no monochromatic complete subgraphs of the given sizes.

The task of finding the actual values of the so called Ramsey numbers $R(b, r)$ is extremely hard in general. There are some estimations but they are far from being sharp.

[14] *Position vectors of coplanar points.* Let the vectors **a**, **b**, **c**, **d** start from the origin, otherwise be arbitrary. If **a**, **b** and **c** are not coplanar then the endpoints of the quadruple **a**, **b**, **c**, **d** are coplanar if and only if there exist real numbers α, β, γ such that

(1) $$\mathbf{d} = \alpha\mathbf{a} + \beta\mathbf{b} + \gamma\mathbf{c}, \qquad \alpha + \beta + \gamma = 1.$$

Indeed, if the endpoints of the vectors are coplanar then the vectors $\mathbf{a} \perp \mathbf{d}$, $\mathbf{b} \perp \mathbf{d}$, $\mathbf{c} \perp \mathbf{d}$ are lying in the same plane and, accordingly, there exist real numbers λ, γ, ν not all zero such that

$$\lambda(\mathbf{a} \perp \mathbf{d}) + \gamma(\mathbf{b} \perp \mathbf{d}) + \nu(\mathbf{c} \perp \mathbf{d}) = \mathbf{0}.$$

Hence

$$(\lambda + \mu + \nu)\mathbf{d} = \lambda\mathbf{a} + \mu\mathbf{b} + \nu\mathbf{c}.$$

Since **a**, **b**, **c** are not coplanar, $\lambda + \mu + \nu \neq 0$, and thus with

$$\alpha = \frac{\lambda}{\lambda + \mu + \nu}, \qquad \beta = \frac{\mu}{\lambda + \mu + \nu}, \qquad \gamma = \frac{\nu}{\lambda_\mu + \nu}$$

we get the desired result. Since the steps of the argument can be reversed the proof is complete.

[15] *Circumscribed parallelepiped.* The endpoints of the non parallel diagonals of two opposite (parallel) faces of a parallelepiped fom a tetrahedron. The edges of this tetrahedron are the face diagonals of the prism and opposite edges are lying on oppposite faces. The prism itself is the circumscribed parallelepiped of the tetrahedron. (*Figure 15.1*)

One can, in fact, construct a circumscribed parallelepiped about any tetrahedron by laying a plane through each edge parallel to the opposite edge; the circumscribed parallelepiped of the tetrahedron is bounded by the planes hence obtained.

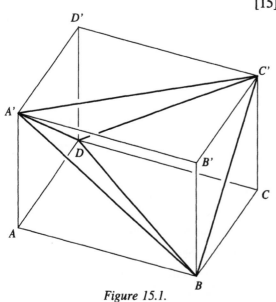

Figure 15.1.

Certain properties of the tetrahedron become transparent when one takes the circumscribed parallelepiped into account. The centroid of a tetrahedron, for example, is incident to the centre of this parallelepiped. The circumscribed parallelepiped of an equilateral tetrahedron is a cuboid, since as the opposite edges of the tetrahedron, the diagonals are equal on each face.

6] *Hero's formula.* Hero's formula expresses the area of a triangle in terms of its sides. There are several ways to draft it:

$$A^2 = s(s \perp a)(s \perp b)(s \perp c) = \frac{1}{16}(a+b+c)(\perp a + b + c)(a \perp b + c)(a + b \perp c) =$$

$$= \frac{1}{16}\left[(a+b)^2 \perp c^2\right]\left[c^2 \perp (a \perp b)^2\right] = \frac{1}{16}(2a^2b^2 + 2b^2c^2 + 2c^2a^2 \perp a^4 \perp b^4 \perp c^4) =$$

$$= \frac{1}{16}\begin{vmatrix} 0 & 1 & 1 & 1 \\ 1 & 0 & c^2 & b^2 \\ 1 & c^2 & 0 & a^2 \\ 1 & b^2 & a^2 & 0 \end{vmatrix}.$$

7] *Specific points of an equilateral tetrhedron*   see [11].

8] *Convex hull.* The convex hull of a set S of points is the convex set containing S and contained by each convex set that contains S. To put it simply the convex hull is the "smallest" convex set containing S,

There is a nice way to visualize the convex hull of a finite planar set: imagine that the points are nails driven in a table and stretch an elastic ribbon about them: the convex hull is the region bounded by the elastic.

Convex hulls in the plane (in the space) can be obtained as the intersections of halfplanes (halfspaces) containing the set S.

9] *Tangent segments in a triangle.* A frequently used fact in elementary arguments that the points of contacts of the circles that are touching the sides of a triangle are dividing them into segments whose lengths can be expressed in a simple manner in terms of the sides of the triangle. The underlying elementary fact is that the tangents to a circle from an external point are congruent. The review of their respective lengths can be checked on the *Figures 19.1* and *19.2*.

0] *Orthocentric tetrahedron.* The lines from the vertices of a tetrahedron perpendicular to the opposite faces are the altitudes of the tetrahedron. If they meet at a common point then this point is called the orthocentre of the tetrahedron and the tetrahedron itself is orthocentric.

Here there are but a few properties of orthocentric tetrahedra:

a) the opposite edges are perpendicular;

b) the sum of the squares is the same for each pair of opposite edges;

c) the faces of their circumscribed parallelepiped are rhombs;

d) their centroid, orthocentre and the centre of the circumscribed sphere are collinear (Euler line).

Conditions a), b), c) are also sufficient for a tetrahedron to be orthocentric; moreover, it is enough to assume that the condition in either a) or b) holds for two opposite pairs of edges only.

[21] *Euler's totient function; Euler's congruence theorem.* The number of the non negative integers up to m that are coprime to m is is the totient function of Euler; its value is denoted by $\varphi(m)$. Here there are a few of its important properties:

1. $a^{\varphi(m)} \equiv 1 \pmod{m}$ if a is a positive integer and $(a, m) = 1$. This is Euler's congruence theorem.

2. If $(a, b) = 1$ then $\varphi(ab) = \varphi(a) \cdot \varphi(b)$.

3. If m is written as the product primes: $m = p_1^{\alpha_1} p_2^{\alpha_2} \ldots p_r^{\alpha_r}$ then,

$$\varphi(m) = m \left(1 \perp \frac{1}{p_1}\right)\left(1 \perp \frac{1}{p_2}\right)\ldots\left(1 \perp \frac{1}{p_r}\right) \quad \text{and } \varphi(1) = 1.$$

[22] *Cauchy's inequality.* Let $(a_1, a_2, \ldots, a_n)$ and $(b_1, b_2 \ldots, b_n)$ be n-tuples of real numbers ("n-dimensional vectors"); then

$$(a_1 b_1 + a_2 b_2 + \ldots + a_n b_n)^2 \leq \left(a_1^2 + a_2^2 + \ldots + a_n^2\right)\left(b_1^2 + b_2^2 + \ldots + b_n^2\right).$$

This inequality is often stated as

$$a_1 b_1 + a_2 b_2 + \ldots + a_n b_n \leq \sqrt{a_1^2 + a_2^2 + \ldots + a_n^2}\sqrt{b_1^2 + b_2^2 + \ldots + b_n^2}.$$

If the numbers b_i are not all zero then equality holds if and only if there exists a real number $\lambda \neq 0$ such that $a_i = \lambda b_i$ $(i = 1, 2, \ldots, n)$ (The n-dimensional vectors are "parallel").

Its simplest proof is as follows: if the numbers b_i are not all zero then the quadratic

$$(a_1 \perp \lambda b_1)^2 + (a_2 \perp \lambda b_2)^2 + \ldots + (a_n + \lambda b_n)^2 = 0$$

in λ has a solution if $a_i = \lambda b_i$ (for every i). Since there can be no more than one solution, its discriminant is not positive and this is exactly the claim. If, on the other hand, the numbers b_i are all zero, then the inequality is obvious.

[23] *Groups.* One of the most frequently occuring algebraic structures. A set of elements forms a group if there is a law of composition which when acting on arbitrary two elements on a definite order assigns an element of the set to this pair. This operation is usually referred to as group multiplication and it is denoted as the ordinary multiplication of numbers. If, for example, a and b are two elements of the group then ab is their product. This operation has the following properties:

1. If a, b, c are the elements of the group then $(ab)c = a(bc)$ (associativiy);

2. There exists an element e of the group such that for any element a

$$ea = ae = a.$$

(e is the neutral element of the group);

3. For any element a of the group there exists an element denoted by a^{-1} such that

$$aa^{-1} = a^{-1}a = e.$$

a^{-1} is the inverse of a.

If $ab = ba$ holds for any two elements, the operation is commutative, then the group is called abelian and the binary operation is then called – and denoted – addition.

Examples

1. The set of integers under addition. Zero is the neutral element and the inverse of each integer is its opposite.

2. The set of real numbers when zero is excluded under multiplication. The neutral element is 1 and the inverse of every element is its reciprocal.

3. The rotations mapping a regular hexagon into itself. The operation is the composition of rotations. The neutral element is the identity transformation and the inverse of the rotation by α is the rotation by $\bot\alpha$. This is a finite non abelian group.

24] *Ptolemy's theorem.* In its general form it states that for the opposite sides a, c and b, d and the diagonals e, f of a convex quadrilateral

$$ac + bd \geq ef,$$

and equality holds if and only if the quadrilateral is cyclic. In the proof we shall refer to the notations of *Figure 24.1*.

Apply, for the triangle DAB a rotation and enlargement about D which maps it into the triangle DCB'. The scale factor of the enlargement is $\dfrac{c}{d}$ and thus $DB' = \dfrac{ec}{d}$ and $CB' = \dfrac{ac}{d}$. The triangles ADC and BDB' are similar because they are mathcing in their angle at D and also in the ratio of the neighbouring sides; the scale factor of this similarity is $\dfrac{e}{d}$ and hence $BB' = \dfrac{f \cdot e}{d}$.

The triangle inequality for the triangle BCB' yields

$$b + \frac{ac}{d} \geq \frac{ef}{d}, \qquad \text{that is} \qquad ac + bd \geq ef.$$

Equality holds if and only if C is lying on the segment BB' that is $\alpha + \gamma = 180°$, the quadrilateral is indeed cyclic.

25] *Equilateral cones.* A circular cone is called equilateral if it has three pairwise perpendicular generators.

We are going to prove the following theorem:

If a circular cone has three pairwise perpendicular generators then there are infinitely many such triples, moreover, every generator is the member of such a perpendicular triple.

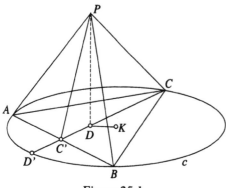

Figure 25.1.

Denote the apex of the cone by P and its base circle by c.

Let the three pairwise perpendicular generators be PA, PB, PC and denote the perpendicular projection of P on the plane of c by D (*Figure 25.1.*). It is easy to prove (Pithagoras, for example) that the triangle ABC is acute angled. Next we show that D is the orthocentre of the triangle ABC. This would follow if the line connecting D to any vertex of the triangle is perpendicular to the opposite side. Consider the vertex C, for example. The side AB is perpendicular to PC because the latter is perpendicular to the plane ABP (it is, in fact, perpendicular to two straight lines lying in this plane) and thus AB is perpendicular to every straight line in the plane ABP. On the other hand, AB is perpendicular to PD since the latter is perpendicular to the plane of c. Therefore, AB is perpendicular to two intersecting lines both lying in the plane PDC and thus it makes a right angle with any line in this plane, CD in particular, therefore, CD is an altitude, indeed.

Denote the intersection of CD with AB by C' and its second intersection with the circle c by D'. Since the mirror image of the orthocentre in a side is incident to the circumcircle, $DC' = C'D'$. In the right triangle $C'PC$ the altitude to the hyptenuse is $PD = h$ and thus, by the geometric mean theorem

(1)
$$PD^2 = h^2 = C'D \cdot DC = \frac{1}{2}D'D \cdot DC.$$

The product $DD' \cdot DC$ is the power of the point D with respect to the circle c. (See [6].) This product is hence independent of C's choice.

Choose now an arbitrary point C_1 on the circle and let C_1D meet the circle at D_1. The perpendicular bisector of DD_1 meets the circle at A_1 and B_1 and the midpoint of DD_1 is C_1'. We show that the generators PA_1, PB_1 and PC_1 are pairwise perpendicular. Since

$$C'D \cdot DC = C_1D \cdot DC_1,$$
$$PD^2 = C_1'D \cdot DC_1,$$

by (1), the triangle $C_1'PC_1$ is right angled (by the converse of the geometric mean theorem). Since A_1B_1 is perpendicular to both C_1D and PD, it is also perpendicular to their plane and hence to PC_1 and thus, by the same way, since PC_1 is perpendicular to PC'_1, it is also perpendicular to PA_1 and PB_1.

We are left to prove that PA_1 and PB_1 are also perpendicular. Note first, for the proof, that D has to be the orthocentre of the triangle $A_1B_1C_1$. Indeed, the orthocentre is the only one point on the altitude from C_1 whose mirror image in the side is lying on the circumcircle and by the construction of the points A_1B_1 the point D now has this property. Similarly, we obtain that PA_1 is perpendicular to PC_1 and also to PB_1. We have thus shown that any generator of the cone belongs to some family of pairwise perpendicular triple of generators.

Circular cones of this property are called equilateral. It follows from the proof that for a given circle c and point P whose distance from the plane of the circle is h and whose perpendicular projection on the base plane is D the point P is the apex of an equilateral cone of directrix c if and only if PD^2 is equal to the half of the power of D with respect to c.

In order to prove that in problem No. 2 of 1978 there are in fact infinitely many cuboids whose vertex opposite to P is Q we have to show that for a given point P and a circle c the vectors $\vec{PA}$, $\vec{PB}$, $\vec{PC}$ are spanning a cuboid whose vertex opposite to P is Q. This would follow had we shown that the sum $\vec{PA} + \vec{PB} + \vec{PC}$ is constant.

Let the vector from the centre K of c to D be $\mathbf{d}$ (this vector is clearly constant). Now

$$\vec{PA} + \vec{PB} + \vec{PC} = 3\vec{PD} + \vec{DA} + \vec{DB} + \vec{DC} =$$
$$= 3\vec{PD} + (\vec{KA} \perp \mathbf{d}) + (\vec{KB} \perp \mathbf{d}) + (\vec{KC} \perp \mathbf{d}) =$$
$$= 3\vec{PD} \perp 3\mathbf{d} + (\vec{KA} + \vec{KB} + \vec{KC}).$$

It is well known that the sum of the vectors from the circumcentre to the vertices of a triangle is leading to the orthocentre. Therefore, $\vec{KA} + \vec{KB} + \vec{KC} = \mathbf{d}$ and thus

$$\vec{PA} + \vec{PB} + \vec{PC} = 3\vec{PD} \perp 2\mathbf{d},$$

and this sum does not depend on the choice of the points A, B, C indeed.

26] *A representation of positive integers.* In the solution of the problem No. 3. of 1978 we have used the following theorem: if α and β are positive irrational numbers satisfying $\dfrac{1}{\alpha} + \dfrac{1}{\beta} = 1$ then the sequences

$$\{[n\alpha]\}, \qquad [\{n\beta\}] \qquad n = 1, 2, \ldots$$

have no common elements and together they exhibit every positive integer.

Note first that the two numbers α and β are greater than 1 and thus the sequences $[n\alpha]$, $[n\beta]$ are strictly increasing. We now prove that, for any positive integer N there is either a positive integer k such that $[k\alpha] = N$, or a positive integer m such that $[n\beta] = N$, moreover, the two options cannot hold at the same time. There clearly exist the unique positive integers k and m such that

$$[(k \perp 1)\alpha] < N \leq [k\alpha], \qquad \text{and}$$
$$[(m \perp 1)\beta)] < N \leq [m\beta].$$

Therefore,

$$k\alpha \perp \alpha < N < k\alpha,$$
$$m\beta \perp \beta < N < m\beta.$$

Subtracting N from each term of the above inequalities:

$$(k\alpha \perp N) \perp \alpha < 0 < k\alpha \perp N,$$
$$(m\beta \perp N) < 0 < m\beta \perp N.$$

Introducing the notations $d = k\alpha \perp N$ and $d' = m\beta \perp N$ we get

(1) $\qquad 0 < d < \alpha, \quad 0 < d' < \beta, \quad \text{that is} \quad 0 < \dfrac{d}{\alpha} < 1, \quad 0 < \dfrac{d'}{\beta} < 1.$

Since $k = \dfrac{N}{\alpha} + \dfrac{d}{\alpha}$, $m = \dfrac{N}{\beta} + \dfrac{d'}{\beta}$, (1) implies

$$k + m = N\left(\frac{1}{\alpha} + \frac{1}{\beta}\right) + \frac{d}{\alpha} + \frac{d'}{\beta} = N + \frac{d}{\alpha} + \frac{d'}{\beta},$$

that is

(2) $\qquad\qquad\qquad \dfrac{d}{\alpha} + \dfrac{d'}{\beta} = k + m \perp N.$

By (1), on the other hand

$$0 < k + m \perp N < 2.$$

Since k, m, N are positive integers, this implies

$$k + m \perp N = 1.$$

Now by (2)

$$\frac{d}{\alpha} + \frac{d'}{\beta} = 1 = \frac{1}{\alpha} + \frac{1}{\beta},$$

that is $\qquad\qquad\qquad \alpha(d' \perp 1) = (1 \perp d)\beta.$

Since α and β are positive and d is irrational, this equality implies that one of d and d' is less than 1 and the other one is greater than 1. If, for example, $d < 1$, $d' > 1$ then, by the definition of d and d' implies

$$\alpha k = N + d, \qquad\qquad [\alpha k] = N,$$
$$\beta m = N + d' \qquad\qquad [\beta m] > N,$$

and thus N belongs to exactly one of the sequences $[\alpha k]$ and $[\beta m]$. The same holds if $d > 1$ and $d' < 1$.

■] *Solving linear recurrences.* To find a formula for the nth term of a sequence defined by recurrence relations is always an important task. Here we present the solution of the following special case of the problem.

(1) $$a_n = c_1 a_{n-1} + c_2 a_{n-2}$$

This is a so called second order linear recurrence, the coefficients a_1 and a_2 are given numbers.

The heart of the matter is to find geometric progressions satisfying (1) and unfold the general solution as a linear combination of these particular sequences. Here there is the method. The quadratic

$$x^2 \perp c_1 x \perp c_2 = 0$$

is called the *characteristic equation* to the recurrence (1). Denote its roots (real or complex) by x_1 and x_2.

Assume, first, that $x_1 \neq x_2$. Then the nth term of the sequence is equal to

(2) $$a_n = \lambda x_1^n + \mu x_2^n,$$

where λ and μ are constants depending on the initial terms of the sequence; their actual value can be computed by solving the simultaneous system

(3)
$$\lambda x_1 + \mu x_2 = a_1,$$
$$\lambda x_1^2 + \mu x_2^2 = a_2.$$

If $x_1 = x_2$ then the nth term can be computed as

$$a_n = \lambda x_0^2 + \mu n x_0^{n-1}$$

(we have adopted the notation $x_1 = x_2 = x_0$). The system for the values of λ and μ is now

$$\lambda x_0 + \mu = a_1,$$
$$\lambda x_0^2 + 2\mu x_0 = a_2.$$

The method works essentially the same way in the general case. Consider the sequence $\{a_i\}$ defined by

(4) $$a_{n+k} = c_1 a_{n+k+1} + c_2 a_{n+k+2} + \ldots + c_k a_n.$$

This relation is called k-order linear recurrence with constant coefficients. The numbers c_i are constants and there are also given the initial values $a_1, a_2, \ldots, a_k$. The characteristic polynomial corresponding to the above recurrence is

$$x^k \perp c_1 x^{k-1} \perp c_2 x^{k-2} \perp \ldots \perp c_k = 0.$$

If its roots (real or complex) are $x_1, x_2, \ldots, x_k$ are distinct then the terms of the sequence $\{a_i\}$ can be computed as

(5) $$a_n = \lambda_1 x_1^{n-1} + \lambda_2 x_2^{n-1} + \ldots + \lambda_n x_k^{n-1}$$

where the coefficients $\lambda_1, \lambda_2, \ldots, \lambda_k$ can be calculated from the initial values of the sequence.

The method still works if there happen to be multiple roots of the characteristic polynomial, although, as in the second order case, the actual solution is a bit more tedious.

[28] *Two relations among binomial coefficients.*

A) $\dbinom{n}{k} = \dbinom{n-1}{k} + \dbinom{n-1}{k-1}$;

B) $\dbinom{n+1}{k+1} = \dbinom{n}{k} + \dbinom{n-1}{k} + \ldots + \dbinom{k}{k}$.

The proof of A) is straightforward from the defining equality $\dbinom{n}{k} =$

$$= \frac{n!}{k!(n-k)!}.$$

$$\binom{n-1}{k} + \binom{n-1}{k-1} = \frac{(n-1)!}{k!(n-k-1)!} + \frac{(n-1)!}{(k-1)!(n-k)!} =$$

$$= \frac{(n-1)!(n-k+k)}{k!(n-k)!} = \binom{n}{k}.$$

To prove B) one can use A):

$$\binom{n+1}{k+1} = \binom{n}{k+1} + \binom{n}{k},$$

$$\binom{n}{k+1} = \binom{n-1}{k+1} + \binom{n-1}{k},$$

$$\binom{n-1}{k+1} = \binom{n-2}{k+1} + \binom{n-2}{k},$$

$$\vdots$$

$$\binom{n-(n-k-2)}{k+1} = \binom{k+2}{k+1} = \binom{k+1}{k+1} + \binom{k+1}{k}.$$

Summing the equalities and considering that $\dbinom{k+1}{k+1} = \dbinom{k}{k}$ one arrives to the claim.

[29] *Menelaus' theorem.* Let C_1, A_1 and B_1 be points on the sides AB, BC, CA of the triangle ABC, respectively. These points are collinear if and only if

(1) $$\frac{AC_1}{C_1B} \cdot \frac{BA_1}{A_1C} \cdot \frac{CB_1}{B_1A} = -1.$$

As for the sign of the fractions on the l. h. s. they are positive if the vectors $\overset{\scriptscriptstyle\mapsto}{AC_1}$ and $\overset{\scriptscriptstyle\mapsto}{C_1B}$ are oriented similarly, otherwise they are negative.

Note here that disregarding the orientation of the segments there is 1 on the r. h. s. of (1) and this form of the claim is just necessary for the points A_1, B_1, C_1 to be collinear.

[0] *Residue classes, congruences.* For a given integer $m > 1$ the integers a and b are said to belong to the same residue class "with respect to m", or simply *modulo* m ("mod m", for short) if they give equal remainders when divided by m. Since the possible remainders are 0, 1, 2, $\ldots$, $m \perp 1$, there are m residue classes with respect to m and every integer belongs to exactly one of them, or, putting it differently, every whole number is representing some residue class.

m integers form a so called *complete residue system* mod m if they represent distinct residue classes; together they hence represent every possible remainder mod m.

Two integers clearly belong to the same residue class mod m if their difference is divisible by m. For given integers a and b this is denoted as

$$a \equiv b \quad (\text{mod } p) \qquad \text{or simply} \qquad a \equiv b \quad (m).$$

This is the relation of congruence. Several properties of this relation are resembling to those of equality; here there are a few of them. (For sake of brevity the mod m extension is now omitted.)

1. if $a \equiv b$, akkor $b \equiv a$; $a \equiv a$ for every integer a;

2. if $a \equiv b$ and $b \equiv c$ then $a \equiv c$;

3. if $a \equiv b$ and $c \equiv d$ then

$a + c \equiv b + d$, $a \perp c \equiv b \perp d$, $ac \equiv bd$, $a^n \equiv b^n$, (n is a positive integer);

4. if $ac \equiv bc$ then $a \equiv b \left(\text{mod } \dfrac{m}{d} \right)$, where $d = (c, m)$.

[31] *Designs.* The v element set H is called block design if there exist b subsets in H of k elements each – the blocks – every element of H belongs to exactly r blocks and any two distinct elements of H belong to exactly λ blocks.

The numbers v, b, k, r, λ are the *parameters* of the block design and they are related in several ways. (Obviously $2 \leq k < v$ and $\lambda > 0$). Assign, to the elements of the block design the rows of a matrix of v rows and b columns and its columns to the blocks as follows: a given entry of the matrix is equal to 1 if the element corresponding to its row belongs to the block corresponding to its column; otherwise the entry is equal to zero. The matrix hence obtained is the so called incidence matrix of the block design. There are exactly r copies of 1 in each row and k 1-s in each column. Tallying the 1-s both row and columnwise yields

$$bk = vr.$$

Counting further incidences one gets

$$r(k \perp 1) = \lambda(v \perp 1).$$

To find the proper conditions under which there exists a block design for a given system of parameters is one of the unsolved hard problems of combinatorics. If, for example, $v = b = p^{2\alpha} + p^{\alpha} + 1$, $\lambda = 1$, $k = r = p^{\alpha} + 1$, where p is a prime and α is a positive integer then one can construct the corresponding block design, these are the so called finite projective planes.

[32] *Fermat's congruence theorem (the little "Fermat's theorem")*. For any prime p the number of coprimes to p below p is equal to $p \perp 1$, $\varphi(p) = p \perp 1$ and by Euler's congruence theorem [21]

$$a^{p-1} \equiv 1 \pmod{p}, \qquad \text{if } (a, p) = 1.$$

This is Fermat's congruence theorem. Multiplying both sides by a yields

$$a^{p} \equiv a \pmod{p}.$$

This form of the theorem holds even if a is not prime to p since as a prime, p then divides a.

[33] *Erdős–Mordell inequality*. Denote the distances of a point P of a triangle from the vertices by p, q and r, respectively, and the perpendicular distances of P from the sides by x, y and z, respectively. Then one has the following inequality

$$p + q + r \geq 2(x + y + z)$$

and equality holds if and only if the triangle is equilateral and P is its centre. This is the Erdős–Mordell-inequality and we are going to prove it.

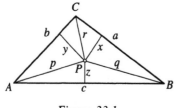

Figure 33.1.

Assume that p, q, r are the distances from the vertices A, B, C and, x, y, z are those from the sides BC, CA, AB, respectively (*Figure 33.1.*).

First we prove the following inequalities:

(1) $ap \geq bz + cy$; $bq \geq cx + az$; $cr \geq zy + bx$.

It is clearly enough to show the first one. Reflect, for the proof, the vertices B and C in the bisector of the angle A and denote the mirror images by C' and B', respectively; the point P remains fixed. The sides of the triangle $AB'C'$ are now $AB' = c$, $B'C' = a$, $C'A = b$; the distances of P from the sides $B'C'$, $C'A$, AB' are x', y, z, respectively (*Figure 33.2.*). The quantity x' refers to signed distance: it is zero, if P is incident to $B'C'$ and if P is outside the triangle $A'B'C'$ then x' is negative.

Denote the altitude from A of the triangle $AB'C'$ by h_a; this is clearly not longer than the path $p + x'$ from A to $B'C'$ through P. Hence

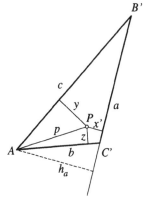

(2) $$m_a \leq p + x'.$$

Multiplying both sides by a and considering that both ah_a and $ax' + bz + cy$ are equal to the double area of the triangle $AB'C'$

$$am_a \leq ap + ax';$$

$$ax' + bz + cy \leq ap + ax',$$

Figure 33.2.

and this implies the claim $ap \geq bz + cy$; a similar argument yields the other two inequalities in (1). Note that the relations hold even if $x' \leq 0$.

Rearranging the inequalities just proved:

$$p \geq \frac{b}{a}z + \frac{c}{a}y;$$

$$q \geq \frac{c}{b}x + \frac{a}{b}z;$$

$$r \geq \frac{a}{c}y + \frac{b}{c}x.$$

The sum of these inequalities gives

(3) $$p + q + r \geq \left(\frac{c}{b} + \frac{b}{c}\right)x + \left(\frac{c}{a} + \frac{a}{c}\right)y + \left(\frac{b}{a} + \frac{a}{b}\right)z.$$

Since the sum of a positive number and its reciprocal is at least 2, one arrives to

$$p + q + r \geq 2(x + y + z).$$

As for equality it must also hold in both (2) and (3). In the latter it holds if and only if $a = b = c$; the sum of a positive number and its reciprocal is 2 only if this number is equal to 1. Our triangle hence must be equilateral.

If this is the case, however, then the triangles ABC and $AB'C'$ in the first paragraph are identical. Hence the position of the segments AP and x' must be identical to that of AP and x; finally, h_a is equal to $(p + x)$ if P is incident to the altitude from A.

This, of course, must hold for the other two altitudes, as well and thus, in case of equality, P must be the intersection of the altitudes, the centre of the equilateral triangle.

[34] *Brocard points.* The points Q_1 and Q_2 are called the Brocard points of the triangle ABC if

$$BAQ_1 \angle = CBA_1 \angle = ACQ_1 \angle, \quad \text{or} \quad ABQ_2 \angle = BCQ_2 \angle = CAQ_2 \angle.$$

These points can be constructed; Q_1, for example, is the intersection of two circles: one of them is passing through A and B and it is touching the line BC and the other one is passing through B and C and touching the line AC. The angles at Q_1 and Q_2 are also equal to each other and

$$\cot \omega = \cot \alpha + \cot \beta + \cot \gamma$$

for their common measure ω. (This is straightforward from the sine rule, for example.)

Apart from he circumcentre of a triangle H the Brocard points are those that have the following property: the feet of the perpendiculars to the sides from these points form a triangle similar to H. This implies that no matter how we rotate the triangle about any one of its Brocard points, the pairwise intersections of the corresponding sides of the two triangles – H and the rotated one – form a triangle that is also similar to H.

[35] *A common origin of certain inequalities.* There is a fundamental inequality showing up in the solutions from time to time. It has various formulations and there are several further inequalities that can be deduced from it. It states that

let $a_1, a_2, \ldots, a_n$ and $b_1, b_2, \ldots, b_n$ are real numbers and $b_{i_1}, b_{i_2}, \ldots, b_{i_n}$ is an arbitrary rearrangement of the numbers b_i. Prepare the sum

$$S = a_1 b_{i_1} + a_2 b_{i_2} + \ldots + a_k b_{i_k}.$$

This sum is maximal if and only if the ordering of the numbers a_i and b_{i_k} is the same and it is minimal if the two orderings are opposite.

Denote, for he proof, the highest terms of the two n-tuples by a_r and b_s, respectively and consider the sum

$$Q = a_1 b_1 + \ldots + a_r b_r + \ldots + a_s b_s + \ldots + a_n b_n.$$

Swap now the two numbers b_r and b_s; if $r \neq s$ then

$$Q' = a_1 b_1 + \ldots + a_r b_s + \ldots + a_s b_r + \ldots + a_n b_n.$$

$$Q \perp Q' = a_r b_s + a_s b_r \perp a_r b_r \perp a_s b_s = (a_r \perp a_s)(b_s \perp b_r) \geq 0.$$

$Q' = Q$ holds if and only if $a_r = a_s$ or $b_r = b_s$, but then the highest a_i is, in fact, multiplied by the highest b_i. Through an appropriate sequence of swaps one arrives to similarly ordered n-tuples. Since the sum Q is not decreasing, the maximum is attained when the orderings are the same, indeed. The corresponding statement about the minimum follows similarly.

6] *Four circles theorem.* Consider four straight lines whose pairwise intersections are distinct. The circumcircles of the four triangles hence obtained are passing through a common point S (*Figure 36.1.*)

Denote, as in the diagram, the intersection of the circumcircles of the triangles ABC and CDE by S. It is clearly enough to show, by symmerty, that the circumcircle of the triangle ADF is, in fact, passing through S. Intercepted by the same arcs $SDE\angle = SCE\angle$ and in the cyclic quadrilateral $ABCS$ this angle is equal to $SAF\angle$, the quadrilateral $ASDF$ is cyclic and thus the circumcircle of the triangle ADF is passing through S, indeed.

By the Simson theorem the feet of the perpendiculars from S to the four lines are collinear. Of what we know about conic sections, it follows that four lines as tangents to the curve determine a unique parabola; the feet of the perpendiculars from the focus to the tangents are on the tangent through the vertex; the circumcircles of the triangles formed by three tangents to the parabola are passing through the focus. Taking these facts into account we get that the point S is, in fact, the focus of the parabola determined by the four straight lines.

7] *Radius inequality.* The diameter of the incircle of a triangle cannot exceed the circumradius, that is

$$R \geq 2r.$$

This is an immediate consequence of Euler's identity $d^2 = R^2 \perp 2Rr$ (d is the distance of the two centres), but there are several proofs around. It is also related to various triangle inequalities and relations, for example

$$\cos \alpha + \cos \beta + \cos \gamma \leq \frac{3}{2},$$
$$\cos \alpha + \cos \beta + \cos \gamma = 1 + \frac{r}{R},$$
$$(\perp a + b + c)(a \perp b + c)(a + b \perp c) \leq abc.$$

The equality $R = 2r$ holds only in equilateral triangles.

8] *Parallel chords.* Consider a circle about the origin on the Argand diagram and denote its four points by the complex numbers a, b, c and d. The chords connecting a to b and c to d are parallel if

$$ab = cd.$$

Assume that the counterclockwise order of the points is a, b, c and d. The corresponding chords are then parallel if and only if the arcs $\overarc{b,c}$ and $\overarc{d,a}$ are equal. Denote the central angle of these arcs by φ and let $e = \cos \varphi + i \sin \varphi$. Multiplication by e is hence a rotation by φ about the origin.

$$be = c \quad \text{and} \quad de = a,$$

from which

$$\frac{be}{de} = ca, \qquad \frac{b}{d} = \frac{c}{a}, \qquad \text{that is} \qquad ab = cd,$$

and the converse of the argument is also valid.

[39] *n-dimensional vectors.* Ordered n-tuples $(a_1, a_2, \ldots, a_n)$ of real numbers are sometimes called n-dimensional vectors and they are then denoted a single bold face letter: $\mathbf{a}(a_1, a_2, \ldots, a_n)$; the numbers a_i are then the coordinates of the vector. There is a natural way to perform algebraic operations between n-dimensional vectors as follows:

addition: the sum of the vectors $\mathbf{a}(a_1, a_2, \ldots, a_n)$ and $\mathbf{b}(b_1, b_2, \ldots, b_n)$ is

$$\mathbf{a} + \mathbf{b}(a_1 + b_1, a_2 + b_2, \ldots, a_n + b_n).$$

Subtraction: $\mathbf{a} \perp \mathbf{b}(a_1 \perp b_1, a_2 \perp b_2, \ldots, a_n \perp b_n).$

Multiplication by the real number λ: $\lambda\mathbf{a}(\lambda a_1, \lambda a_2, \ldots, \lambda a_n).$

Scalar, or dot product: $\mathbf{ab} = a_1 b_1 + a_2 b_2 + \ldots + a_n b_n.$

The fundamental algebraic laws are as follows:

$$\mathbf{a} + \mathbf{b} = \mathbf{b} + \mathbf{a}, \qquad \mathbf{ab} = \mathbf{ba}, \qquad \lambda\mathbf{a} = \mathbf{a}\lambda, \qquad \lambda(\mu\mathbf{a}) = (\lambda\mu)\mathbf{a} = \lambda\mu\mathbf{a},$$

$$(\lambda + \mu)\mathbf{a} = \lambda\mathbf{a} + \mu\mathbf{a}, \qquad \lambda(\mathbf{a} + \mathbf{b}) = \lambda\mathbf{a} + \lambda\mathbf{b}, \qquad \lambda(\mathbf{ab}) = (\lambda\mathbf{a})\mathbf{b} = \mathbf{a}(\lambda\mathbf{b}),$$

$$\mathbf{a}(\mathbf{b} + \mathbf{c}) = (\mathbf{b} + \mathbf{c})\mathbf{a} = \mathbf{ab} + \mathbf{ac} \qquad \text{(distributive law).}$$

Further notations and concepts: $\dfrac{\mathbf{a}}{\lambda} = \dfrac{1}{\lambda}\mathbf{a}$. The vector $\mathbf{0}(0, 0, \ldots, 0)$ is called zero vector; the product of two equal vectors is called the square of the given vector and abbreviated accordingly: $\mathbf{aa} = \mathbf{a}^2 = a_1^2 + a_2^2 + \ldots + a_n^2.$

The distributive law also holds if both factors have several terms; in particular:

$$(\mathbf{a}_1 + \mathbf{a}_2 + \ldots + \mathbf{a}_n)^2 = \mathbf{a}_1^2 + \mathbf{a}_2^2 + \ldots + \mathbf{a}_n^2 + 2(\mathbf{a}_1\mathbf{a}_2 + \mathbf{a}_1\mathbf{a}_3 + \ldots + \mathbf{a}_{n-1}\mathbf{a}_n).$$

The proof of any one of the above relations can be done by expanding them in terms of coordinates. For a further application see also [22].

[40] *Weighed means.* The notion of weighed means is a generalization of the notion of means.

Assign, as its weight, to each of the real numbers $a_1, a_2, \ldots, a_n$ a positive number s_i, respectively. The weighed arithmetic mean (or weighed average) of the numbers $a_1, a_2, \ldots, a_n$ is then

$$A_s = \frac{s_1 a_1 + s_2 a_2 + \ldots + s_n a_n}{s_1 + s_2 + \ldots + s_n};$$

their weighed geometric mean is

$$G_s = \sqrt[s_1 + s_2 + \ldots + s_n]{a_1^{s_1} a_2^{s_2} \ldots a_n^{s_n}};$$

the weighed harmonic mean is

$$H_s = \frac{s_1 + s_2 + \ldots + s_n}{\frac{s_1}{a_1} + \frac{s_2}{a_2} + \ldots + \frac{s_n}{a_n}},$$

and, finally, the weighed quadratic mean is

$$Q_s = i\frac{s_1 a_1^2 + s_2 a_2^2 + \ldots + s_n a_n^2}{s_1 + s_2 + \ldots + s_n}.$$

Togeteher these weighed means also obey the well known chain of inequalities between ordinary means, namely

$$H_s \leq G_s \leq A_s \leq Q_s.$$

The proof follows a general pattern. The first step it is straightforward: if the weights s_i are whole numbers then there is nothing new here, the weighed means can be conceived as ordinary means of appropriate number of copies of each number: there are s_i occurences of the number a_i. If the weights are rational then, as the following example shows, the issue can be reduced to the previous case. Let's see how to do this in the $A_s \geq G_s$ inequality for two terms; let the weights be $s_1 = \frac{p_1}{q_1}$ and $s_2 = \frac{p_2}{q_2}$ (p_i and q_i are positive integers). Then

$$A_s = \frac{\frac{p_1}{q_1}a_1 + \frac{p_2}{q_2}a_2}{\frac{p_1}{q_1} + \frac{p_2}{q_2}} = \frac{p_1 q_2 a_1 + p_2 q_1 a_2}{p_1 q_2 + p_2 q_1} \geq {}^{p_1 q_2 + p_2 q_1}\sqrt{a_1^{p_1 q_2} \cdot a_2^{p_2 q_1}} =$$

$$= {}^{\frac{p_1}{q_1} + \frac{p_2}{q_2}}\sqrt{a_1^{\frac{p_1}{q_1}} \cdot a_2^{\frac{p_2}{q_2}}} = G_s.$$

Finally, if there happen to be irrational numbers among the weights, then one should invoke standard continuity arguments. The point is that the means are continuous functions of the weights and irrational numbers can be approximated to arbitrary precision by rationals.

41] *Trigonometric form of Ceva's theorem and an application.* If the lines a', b', c' are dividing the angles α, β, γ of the triangle ABC into the parts α_1 and α_2, β_1 and β_2, γ_1 and γ_2, respectively (*Figure 41.1*) then the lines a', b', c' are concurrent if and only if

(1)
$$\frac{\sin \alpha_1 \sin \beta_1 \sin \gamma_1}{\sin \alpha_2 \sin \beta_2 \sin \gamma_2} = 1.$$

Assume first that the three lines in question are passing through a common point P. By the sine rule in the triangles ABP, BCP, CAP respectively

$$\frac{PA}{PB} = \frac{\sin \beta_1}{\sin \alpha_2}, \qquad \frac{PB}{PC} = \frac{\sin \gamma_1}{\sin \beta_2},$$

$$\frac{PC}{PA} = \frac{\sin \alpha_1}{\sin \gamma_2}.$$

and the product of the three equalities yields (1).

Assume, for the converse, that the lines a', b', c' divide the angles of the triangle according to (1). Denote the intersection of the lines a' and b' by P' and suppose that $P'C$ cuts the angle γ into the parts γ_1' and γ_2'. We have already seen that

$$\frac{\sin\alpha_1 \sin\beta_1 \sin\gamma_1'}{\sin\alpha_2 \sin\beta_2 \sin\gamma_2'} = 1,$$

which, when compared to (1), yields $\dfrac{\sin\gamma_1}{\sin\gamma_2} = \dfrac{\sin\gamma_1'}{\sin\gamma_2'}$, that is

$$\frac{\sin(\gamma \perp \gamma_2)}{\sin\gamma_2} = \frac{\sin(\gamma \perp \gamma_2')}{\sin\gamma_2'}, \quad \text{or}$$

$$\sin\gamma_2'(\sin\gamma\cos\gamma_2 \perp \cos\gamma\sin\gamma_2) = \sin\gamma_2(\sin\gamma\cos\gamma_2' \perp \cos\gamma\sin\gamma_2'),$$

$$\sin\gamma_2'\cos\gamma_2 = \sin\gamma_2\cos\gamma_2',$$

$$\sin(\gamma_2' \perp \gamma_2) = 0.$$

This implies $\gamma_2 = \gamma_2'$ and $\gamma_1 = \gamma_1'$ and thus P' and P are identical, the proof is complete.

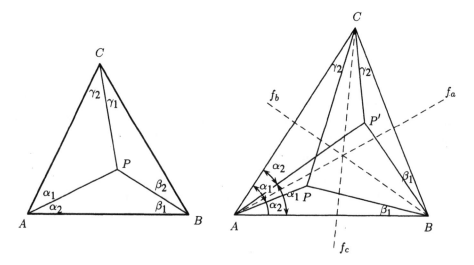

Figure 41.1.   *Figure 41.2.*

An immediate consequence is the theorem used in the solution of Qu. No 2. in 1996: if P is an interior point of the triangle ABC and the line PA is reflected in the bisector of $A\angle$, PB is reflected to the bisector of $B\angle$, finally PC is reflected in the bisector of $C\angle$, then the reflected lines are concurrent. Indeed, the parts α_1 and α_2, β_1 and β_2, γ_1 and γ_2 are swapped under the reflections and hence (1) remains valid, the mirror lines are also passing through a common point (*Figure 41.2.*).

This assertion can be proved in a more general form without the Ceva-trigonometry machinery using reflections only; if the lines through the vertices of a triangle belong to a pencil (i.e. they are concurrent or parallel) then the same holds for the mirror images in the corresponding angle bisectors.

2] *An extension of the Erdős–Mordell inequality.* Let P, Q and S be interior points on the sides AB, BC, CA of the triangle ABC, respectively. Denote the intersection of the perpendiculars to AB at P and to BC at Q by Y and similarly, the intersection of the perpendiculars to BC at Q and to CA at S by Z and, finally, the intersection of the perpendiculars to CA at Q and to AB at P by X. If, additionally, the points X, Y and Z are interior to the triangle ABC then

$$AX + BY + CZ \geq XP + YP + YQ + ZQ + ZS + XS,$$

and equality holds if and only if the triangle ABC is equilateral and the each of the points X, Y, Z are at the centre of ABC. (*Figure 42.1.*).

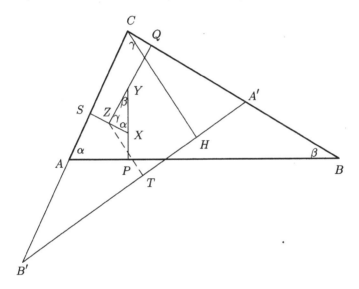

Figure 42.1.

The proof is following the demonstration in [33] of the original theorem. Denote by A', B' the mirror image of the vertices A and B in the interior bisector of the angle C, respectively; the feet of the perpendiculars to $A'B'$ from C and Z be H and T, respectively. The area of the triangle $A'B'C$ can be written in two different ways:

$$2a_{A'B'C} = 2a_{A'B'Z} + 2a_{CA'Z} + 2a_{B'CZ}.$$

With the compulsory notations $AB = A'B' = c$, $BC = B'C = a$, $CA = CA' = b$ this can be put as

(1) $$c \cdot CH = c \cdot ZT + b \cdot ZQ + a \cdot ZS;$$

here the length of ZT is negative if the line $A'B'$ separates the points Z and C. In any case we have the following inequality:

(∗) $$CZ + ZT \geq CH.$$

Hence

$$c \cdot CZ + c \cdot ZT \geq c \cdot CH,$$

$$c \cdot CZ \geq c \cdot CH \perp c \cdot ZT,$$

which, when combined with (1), implies

$$c \cdot CZ \geq b \cdot ZQ + a \cdot ZS,$$

$$CZ \geq \frac{b}{c} \cdot ZQ + \frac{a}{c} ZS.$$

Similarly

$$AX \geq \frac{c}{a} XS + \frac{b}{a} XP,$$

$$BY \geq \frac{a}{b} YP + \frac{c}{b} YQ.$$

Adding these inequalities

(2) $$AX + BY + CZ \geq \left(\frac{a}{b}YP + \frac{b}{a}XP\right) + \left(\frac{b}{c}ZQ + \frac{c}{b}YQ\right) + \left(\frac{c}{a}XS + \frac{a}{c}ZS\right).$$

Apply now the identity

$$kK + nN = (k+n)\frac{K+N}{2} + (k \perp n)\frac{K \perp N}{2}$$

for the expression

$$\frac{a}{b}YP + \frac{b}{a}XP = \left(\frac{a}{b}+\frac{b}{a}\right)\frac{YP+XP}{2} + \left(\frac{a}{b}\perp\frac{b}{a}\right)\frac{YP\perp XP}{2}.$$

Observe that their angles being pairwise equal the triangles ABC and XYZ are similar. If λ is the scale factor of similarity then

$$\frac{YZ}{a} = \frac{ZX}{b} = \frac{XY}{c} = \lambda,$$

and also $YP \perp XP = XY = \lambda c$. By

(**) $$\frac{a}{b} + \frac{b}{a} \geq 2.$$

we now get

$$\frac{a}{b} \cdot YP + \frac{b}{a} \cdot XP \geq YP + XP + \lambda\left(\frac{ca}{2b} \perp \frac{bc}{2a}\right).$$

Similarly

$$\frac{b}{c} \cdot ZQ + \frac{c}{b} YQ \geq ZQ + YQ + \lambda\left(\frac{ab}{2c} \perp \frac{ca}{2b}\right),$$

$$\frac{c}{a} \cdot XS + \frac{a}{c} \cdot ZS \geq XS + ZS + \lambda\left(\frac{bc}{2a} \perp \frac{ab}{2c}\right).$$

By (2) the sum of these inequalities implies the claim.

$$AX + BY + CZ \geq XP + YP + YQ + ZQ + ZS + XS.$$

If there is equality then by (**) $a = b = c$ and by (*) X, Y and Z are laying on the respective altitudes; summarizing the conditions of equality ABC has to be equilateral and each of X, Y, Z must be at its centre.

3] *A property of equilateral triangles.* Let P be an arbitrary point on the arc AB not containing C of the circumcircle of the equilateral triangle ABC. Then $AP + BP = PC$.

Rotate, about A, the triangle APB by $60°$. If the image is ACP' then P' is on the segment CP because $ACP\angle = ABP\angle$ (inscribed angles intercepting the same arc AP). By the rotation the triangle APP' is equilateral and thus $AP = PP'$; since, on the other hand, $BP = CP'$ we get

$$PC = PP' + CP' = AP + BP,$$

indeed. (*Figure 43.1.*).

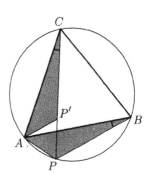

Figure 43.1.

We note that the claim is an immediate consequence of Ptolemy's theorem [24]; when applied to the cyclic quadrilateral $APBC$ it yields

$$AP \cdot BC + BP \cdot CA = AB \cdot PC,$$

and dividing through by the length of the side of the triangle we get the desired result.

4] *The number of divisors.* The number of (positive) divisors of a positive integer n is denoted by $d(n)$. (One can also come across to the notation $\tau(n)$.) If the prime factorization of n is

$$n = p_1^{\alpha_1} \cdot p_2^{\alpha_2} \cdot \ldots \cdot p_r^{\alpha_r},$$

then every divisor of n is equal to

$$p_1^{\beta_1} p_2^{\beta_2} \ldots p_r^{\beta_r}$$

where $0 \leq \beta_i \leq \alpha_i$, and, conversely, for any such choice of the numbers β_i there is a divisor of the above form. Therefore, there are $\alpha_i + 1$ ways to set the value of β_i and, accordingly

$$d(n) = (\alpha_1 + 1)(\alpha_2 + 1) \ldots (\alpha_r + 1)$$

divisors of n, altogether. This shows that $d(n)$ depends on the list of indices only, not on the actual prime factors. As a consequence we note here that if a and b are coprime then

$$d(ab) = d(a)d(b),$$

the function d is multiplicative. This property is, of course, true for products of finitely many coprime factors.

[45] *Turán's graph theorem.* Paul Turán proved the following theorem in 1941: let $n = q(k-1) + r$, where q, k, r are whole numbers such that $0 \le r < k-1$. If there are more than

$$E = \frac{k-2}{2(k-1)}(n^2 - r^2) + \binom{r}{2},$$

edges in a simple graph of n vertices then the graph contains a complete subgraph of k vertices. The result is sharp, since for every n there exists a simple graph of n vertices and E edges with no complete subgraph of k vertices.